International Library of Ethics, Law, and the New Medicine

Volume 80

Series editors
David N. Weisstub, Faculty of Medicine, University of Montreal, Montreal, QC, Canada
Dennis R. Cooley, History, Philosophy, and Religious Studies, North Dakota State University, Fargo, ND, USA

Founded by
Thomasine Kimbrough Kushner, Berkely, USA
David C. Thomasma, Dordrecht, The Netherlands
David N. Weisstub, Montreal, Canada

The book series *International Library of Ethics, Law and the New Medicine* comprises volumes with an international and interdisciplinary focus. The aim of the Series is to publish books on foundational issues in (bio) ethics, law, international health care and medicine. The volumes that have already appeared in this series address aspects of aging, mental health, AIDS, preventive medicine, bioethics and many other current topics. This Series was conceived against the background of increasing globalization and interdependency of the world's cultures and governments, with mutual influencing occurring throughout the world in all fields, most surely in health care and its delivery. By means of this Series we aim to contribute and cooperate to meet the challenge of our time: how to aim human technology to good human ends, how to deal with changed values in the areas of religion, society, culture and the self-definition of human persons, and how to formulate a new way of thinking, a new ethic. We welcome book proposals representing the broad interest of the interdisciplinary and international focus of the series. We especially welcome proposals that address aspects of 'new medicine', meaning advances in research and clinical health care, with an emphasis on those interventions and alterations that force us to re-examine foundational issues.

More information about this series at http://www.springer.com/series/6224

Wanda Teays

Doctors and Torture

Medicine at the Crossroads

 Springer

Wanda Teays
Philosophy Department
Mount Saint Mary's University
Los Angeles, CA, USA

ISSN 1567-8008   ISSN 2351-955X (electronic)
International Library of Ethics, Law, and the New Medicine
ISBN 978-3-030-22516-2   ISBN 978-3-030-22517-9 (eBook)
https://doi.org/10.1007/978-3-030-22517-9

© Springer Nature Switzerland AG 2019
This work is subject to copyright. All rights are reserved by the Publisher, whether the whole or part of the material is concerned, specifically the rights of translation, reprinting, reuse of illustrations, recitation, broadcasting, reproduction on microfilms or in any other physical way, and transmission or information storage and retrieval, electronic adaptation, computer software, or by similar or dissimilar methodology now known or hereafter developed.
The use of general descriptive names, registered names, trademarks, service marks, etc. in this publication does not imply, even in the absence of a specific statement, that such names are exempt from the relevant protective laws and regulations and therefore free for general use.
The publisher, the authors, and the editors are safe to assume that the advice and information in this book are believed to be true and accurate at the date of publication. Neither the publisher nor the authors or the editors give a warranty, express or implied, with respect to the material contained herein or for any errors or omissions that may have been made. The publisher remains neutral with regard to jurisdictional claims in published maps and institutional affiliations.

This Springer imprint is published by the registered company Springer Nature Switzerland AG.
The registered company address is: Gewerbestrasse 11, 6330 Cham, Switzerland

*To the memory of Woodrow C. Teays
Geronimo navigator and atomic vet*

Foreword[1]

Wanda Teays presents in her book *Doctors and Torture: Medicine at the Crossroads* key ethical and practical issues that are at the center of this public debate. On the one hand, countries sometimes feel threatened, such as the United States, in the wake of the terrorist attack of 9/11/01. No country wants to feel threatened in this way. The "fight or flight" response kicks in and safety may be given so much precedence in the political triage that citizens will support *any* policy that promises to alleviate such a threat. No country *should* feel threatened in this way. All of this Teays concedes. But, on the other hand, the question that comes next is as follows: *What should we do about it?* This refers both to solve the question of who were the perpetrators (essentially a police action) and how should we act to protect ourselves against future immoral reactions to the same.

In Chaps. 2 and 3, Teays creates a scenario of double consciousness where the response of *torture* to these events is compared to *lying*: everyone is uncomfortable with it, but maybe it is necessary. This may be a common civic attitude, but what about the physicians? Do they have a professional duty to rise above this popular reaction? Do doctors have a duty not only to refrain from participating in torture but also to report it when they see it and to protect the evidence? This is the ground of Teays' book.

When we think of the global considerations, one barrier for complete analysis is all the secrecy surrounding torture. It is hard to assess the cause of the moral failure if we do not have all the facts. So what should our reaction be to all this secrecy? Either we double down and investigate (as Italy and the United Kingdom have done) or we resign ourselves and quit. This latter reaction will metastasize the tumor of torture as public policy.

In Chap. 4, Teays helps us visualize torture through six examples. From these, we find certain common features: disregard for the suspect's dignity and reinforcement of the power dynamic *(might makes right)*. This latter result becomes a driving

---

[1] An example of covering up wrongdoing that enabled crimes to continue is in the oversight by bishops and archbishops in the Roman Catholic Church concerning pedophile activity by members of the clergy. Unless there is vigorous oversight, the problem just continues.

force that influences the rules of engagement: various methods of inflicting pain as a way to gather information. This is done naively without empirical backing on whether it is even an efficient method. Instead, the power dynamic creates a community worldview of inflicting long-term brutality.

In Chap. 5, one of the critical chapters of the book, there is an exploration of the effects of torture. Following some distinctions set out by Nel Noddings on evil, Teays applies this to torture, showing that as a result of these crimes, pain, separation, and helplessness occur. These are medical symptoms that our healthcare professionals should respond to with care. Physicians (and all healthcare professionals) need to take the side of the patient and not of the "disease" (the interrogators). But, instead, many physicians took the government's side to assist in the design of the torture. They used their knowledge of mind and body to help create a program of maximum pain, separation, and helplessness. Sometimes, these methods even resulted in death. But the physicians justified their collaboration by placing "loyalty to country above their care for detainees." This is a direct repudiation of their professional duty as physicians.

In Chap. 6, we are presented with factual details of the methods used in these "enhanced interrogations." These methods included forced shaving, prolonged diapering, sexual humiliation, etc. These go against Kant's second form of the categorical imperative that commands that we recognize the dignity of all people (treat them as ends) and never lower them to mere instruments of some social or practical design (treating them as means only). This chapter connects specific activities of torture to the general moral duty that applies to all people regardless of their profession. Thus, medical personnel who support torture are *immoral people* as well as bad doctors.

In Chap. 7, we explore solitary confinement. This is a very controversial form of imprisonment, in general. Even within our civilian population, it is thought by most that solitary confinement constitutes *cruel and unusual punishment.* The source of this is that biologically, we are social animals, so taking this away amounts to attack the detainee at his roots as a human. As Teays points out, there is no instrumental value in treating prisoners of war in this manner. It is done merely to create pain and suffering. Teays notes that Britain and Norway have moved away from the solitary confinement even in their domestic corrections facilities. Teays suggests the creation of independent citizen boards will stop this practice. I agree.

Chapters 8 and 9 move toward general ethical theory and professional ethics, respectively. Now, in the realm of personal and social/political ethics, there are various normative theories that generate their imperatives for action on different grounds. Teays set these out as teleological theories, deontological theories (along with Kantian theories), virtue ethics, and feminist ethics. Sometimes, this leads to arguments on the details and can derail the overall discussion. Teays wisely avoids this quagmire by pointing out the areas of agreement among most moral theories against degradation and dehumanization (sometimes resulting in permanent incapacity or even death). By following up each theory presented with application, the reader is clear on Teays' take on each. (For those considering using this book in the classroom, this is especially useful.)

However, there is one theory that leaves the door ajar here: utilitarianism. If the majority considers the outcome to be worth it, some practitioners of utilitarianism will allow torture.

These practitioners often bring up the *ticking time bomb thought experiment.* In this thought experiment, there is a bomb that will kill a large number of people—many thousands or more. Since the death of one causes less negative happiness than the death of thousands, it seems permissible to kill one to save thousands. And since torture is less extreme than killing, the proponent says that if torture will bring the desired information, then it is ethically permissible.

Teays emphasizes the fact that brutality casts a long shadow. This long shadow confronts those who try to bring forward such examples. In the long run, the choice of torture poisons a society that allows it. Therefore, the utilitarian will have to demur. That only leaves ethical egoists and ethical relativists as proponents. But they, also, fall by extending the time frame of consequences. Teays, again, makes the right call here. With the temporal parameter in play, no ethical theory will condone torture.

Then, there is the issue of professional ethics. The classic code of professional medical ethics goes back to Hippocrates. There are two main tenets: do no harm, and promote the good of the patient. Torture—especially when assisted by medical personnel—violates both of these tenets. It is grounds for losing one's medical license; therefore, *physicians may not be involved in any way in the torture of human beings*—period. Teays brings this essential point of professional ethics home with vigor. She recommends that, first, physicians be held accountable for acts of complicity in torture and, second, that covering up records of malfeasance should be viewed as crime as well. This second point is very important because covering up immoral, criminal activity enables the continuation of that activity.

The convergence of general ethics with professional ethics is a cornerstone of the book and will make this tome valuable for those doing research on this subject and for those looking for a powerful addition to classes in *global ethics*, *just war theory*, and *global medical ethics* (among others).

Teays says it best when she argues that professional standards supersede political ideology. This is often difficult in a politics-driven world, but it is a caution we must accept if we are to *escape* the fate of a "might makes right" world that accepts no values aside from power.

Dr. Wanda Teays' presentation is one crucial argument to help us avoid this fate. Thank you, Wanda.

Marymount University Michael Boylan
Arlington, VA, USA

Acknowledgments

Thank you Floor Oosting, executive editor of *Social Sciences*; Christopher Wilby, assistant editor; and the project coordinators for seeing this book from proposal through production. I am most grateful for your encouragement, your patience, and all you did to make *Doctors and Torture* happen.

Thank you Mount Saint Mary's University for my sabbatical to work on this book. Thanks also to Michael Boylan for writing a foreword to this text and to Willow Bunu for her reference and research skills. Both of you have been a source of support and encouragement. I am so very appreciative.

And thank you dear reader. We have more work ahead, as this book makes clear. Your commitment to examine the concerns around torture and to effect change makes all the difference.

Contents

1	**Introduction**	1
	1.1 Structure of This Book	2
	Works Cited	6

Part I Challenges for the Profession

2	**Torture and Terrorism Overview**	9
	2.1 Introduction	9
	2.2 What is Considered "Terrorism"?	10
	2.3 The Power of Terrorist Rhetoric	11
	2.4 Tools for Framing the Discussion	11
	2.5 How Do We Set Boundaries?	12
	2.6 It Pays to Consider the Labels We Use	13
	2.7 Grappling With the Concept of Evil	13
	2.8 Evil and Torture	14
	2.9 Different Kinds of Warfare	15
	2.10 Polarization in the War of Terror	16
	2.11 Problems With Polarization in the War on Terror	16
	2.12 How Do We Bolster the "Ethics" in Medical Ethics?	18
	2.13 The Path to Torture	18
	2.14 The Terms We Use Make All the Difference	19
	2.15 Doctors Add Legitimacy	20
	2.16 Conflicting Loyalties, Messy Dilemmas	20
	2.17 Doctors' Susceptibility to Pressure	21
	2.18 Torture Casts a Long Shadow	21
	2.19 Conclusion	22
	Works Cited	22
3	**Global Considerations**	25
	3.1 Introduction	26
	3.2 Doctors' Involvement in Torture	27
	3.3 Exceptions Can Become Normalized	28

	3.4	And Doctors Get Pulled In	28
	3.5	Psychologists Cross the Ethical Boundary Too	29
	3.6	From the Few to the Many	30
	3.7	Moral Quandaries in Assisting Torture Victims	30
	3.8	The Extent of Doctors' Participation in Torture	31
	3.9	There Are Various Ways to Lose One's Bearings	32
	3.10	There Are Still Lessons to Learn	32
	3.11	Secrecy	33
	3.12	The Questionable Effectiveness of Torture	34
	3.13	Rendition	34
	3.14	The CIA "Black Sites" Offer the Ultimate Secrecy	35
	3.15	Secrecy Undermines Human Rights	36
	3.16	Conclusion	36
		Works Cited	38
4	**Global Torture**		41
	4.1	Introduction	41
	4.2	First Case: The Mau Mau of Kenya (1950s)	42
	4.3	Doctors Played a Role in the Abuse	42
	4.4	Second Case: Colonia Dignidad Nazi Camp in Chile (1970s)	43
	4.5	Third Case: Torture in the Philippines (2014)	44
	4.6	Torturers and Torture Victims	45
	4.7	Fourth Case: Uzbekistan (2014–2015)	45
	4.8	The Lack of Transparency Adds to the Problem	46
	4.9	Fifth Case: Syria (2003–2017)	46
	4.10	Sixth Case: Abu Ghraib	47
	4.11	Conclusion	49
	4.12	And Then There Is the Medical Profession	50
		Works Cited	51

Part II Boundaries of Torture

5	**Boundaries of Torture**		57
	5.1	Introduction	58
	5.2	Where Do We Find Torture—And By Whom?	58
	5.3	The Infliction of Pain	59
	5.4	The "Five Techniques"	59
	5.5	The Domain of Pain: Torture on the Physical Plane	60
	5.6	The Use of Mental and Psychological Abuse	60
	5.7	Justifying Torture with Hypotheticals	61
	5.8	The Slippery Slope of "Force Drift"	62
	5.9	Putting "Force Drift to Work	62
	5.10	Mental Pain and Torture	63
	5.11	Employing a "Mind Virus"	64
	5.12	Making Victims Feel Responsible for Their Pain	65

Contents

5.13	Clustering: Combining Methods of Abuse	65
5.14	How Does Language Figure in?	66
5.15	Twisting Language	66
5.16	And Then It's All Couched in Secrecy	67
5.17	The Impact of Manipulating Language	67
5.18	The Language Games	68
5.19	Strategies for Justifying "Forceful"/"Enhanced" Interrogation	69
5.20	What Are the Consequences of Such Strategies?	69
5.21	Shaping Public Consciousness	70
5.22	The Use of Language: Medical Personnel	71
5.23	Conclusion	71
Works Cited		72

6 Degradation and Dehumanization 75
  6.1 Introduction 76
  6.2 First Strategy: Hooding 77
  6.3 The Cost Is Not Insignificant 78
  6.4 Second Strategy: Forced Nudity 78
  6.5 Three: Force-Feeding 80
  6.6 Breaking Hunger Strikes with Force-Feeding 81
  6.7 "Rectal Hydration" 81
  6.8 Responding to Hunger Strikes 82
  6.9 What About Other Kinds of Forced Treatment? 83
  6.10 Not All Enable Degrading Practices 83
  6.11 Addressing Dual Loyalties 84
  6.12 Four: Waterboarding 84
  6.13 Five: Humiliation 87
  6.14 Prolonged Diapering 88
  6.15 Another Form of Humiliation Is Sexual Abuse 88
  6.16 The Case of Abu Ghraib 88
  6.17 The Case of Yemeni Prisons 89
  6.18 Conclusion 89
  Works Cited 90

7 Solitary Confinement 95
  7.1 Introduction 96
  7.2 The Misuse of Solitary 96
  7.3 First Case: Lawyer Xie Yanyi 97
  7.4 Second Case: Detainee Jose Padilla 97
  7.5 Third Case: Hostages Terry Anderson and Frank Reed 98
  7.6 Fourth Case: Prisoners Robert King, Herman Wallace,
       and Albert Woodfox, AKA "The Angola Three" 98
  7.7 Fifth Case: Death Row Convicts Marcus Hamilton,
       Winthrop Eaton and Michael Perry 99
  7.8 Overview of Issues 99
  7.9 The Harms of Solitary 100

7.10	Abusive and Aggressive Forms of Solitary	101
7.11	Descriptions and Denials	101
7.12	The Use of Euphemisms	102
7.13	Where Is the Outrage?	103
7.14	Working for Change	103
7.15	What Role Can Doctors Play?	104
7.16	Conclusion	105
Works Cited		106

Part III Ethical Assessment

8 Ethical Theory 111

8.1	Introduction	111
8.2	Moral Agency and Culpability	112
8.3	The Major Ethical Theories	113
8.4	Teleological Ethics	113
8.5	Ethical Egoism	113
8.6	Applying the Theory	114
8.7	Ethical Relativism	115
8.8	Applying the Theory	115
8.9	Utilitarianism	116
8.10	Applying the Theory	117
8.11	Deontological Ethics	118
8.12	Kantian Ethics	119
8.13	Applying the Theory	120
8.14	W.D. Ross's Prima Facie Duties	121
8.15	Applying the Theory	121
8.16	John Rawls's Justice Theory	122
8.17	Applying the Theory	123
8.18	Virtue Ethics	124
8.19	Applying the Theory	125
8.20	Feminist Ethics	125
8.21	Applying the Theory	127
8.22	Conclusion	128
Works Cited		128

9 Applied Ethics: Principles and Perspectives 131

9.1	Introduction		131
9.2	Beauchamp's and Childress's Basic Principles of Western Bioethics		132
	9.2.1	Principle #1: Patient Autonomy	133
	9.2.2	Principles #2 and #3: Beneficence and Non-maleficence	133
	9.2.3	Principle #4: Justice	134
9.3	Torture Is Unjust		134

9.4	Bernard Gert's Universal Moral Rules	135
	9.4.1 Do Not Kill	136
	9.4.2 Do Not Deceive	136
	9.4.3 Do Not Cause Pain	136
	9.4.4 Keep Your Promises	137
	9.4.5 Do Not Disable	137
	9.4.6 Do Not Cheat	137
	9.4.7 Do Not Deprive of Freedom	137
	9.4.8 Obey the Law	137
	9.4.9 Do Not Deprive of Pleasure	138
	9.4.10 Do Your Duty	138
9.5	The World Medical Association	138
9.6	The Declaration of Tokyo	140
9.7	The American Medical Association	141
9.8	The International Council of Nurses	142
9.9	Istanbul Protocol: Obligations to Prevent Torture	143
9.10	Physicians for Human Rights	144
9.11	Conclusion	144
9.12	What, Then, Follows from This?	145
9.13	Working for Change	145
Works Cited		146

Chapter 1
Introduction

> *This is a moral debate. It is about who we are.*
> —Sen. John McCain.

> Torture makes you go mad. Sometimes I catch myself going mad again now. Every time I am force-fed, every time I meet with my lawyer, every time I see a doctor, they use some kind of metal detector device to do a cavity search. They have never found anything in all these years. What I am meant to be hiding, I have no idea. It is pointless. But I have to wonder if the radiation it emits isn't my own private Hiroshima or Nagasaki — four, six, eight times a day. Maybe I am paranoid, but I feel that something bad is happening to me, deep inside (Rabbani 2018).

So says detainee Ahmed Rabbani. He laments, "I'm stuck in Guantanamo. The world has forgotten me." In light of the fact that he's been there for 14 years with no end in sight, it is no wonder he's drawn this conclusion. Rabbani is not alone in fearing that he's now invisible in the eyes of the world.

His plea notwithstanding, far too little attention has been given to the treatment of prisoners and detainees. Far too little concern has been given to their abuse and torture at the hands of guards and interrogators. And far too little assessment has been done on the involvement of medical personnel. Who speaks for the victims? Who cares about their fate?

Consider this second example, that of Khalid Sheikh Mohammed:

> Mr. Mohammed was subjected to the suffocation technique called waterboarding 183 times over 15 sessions, stripped naked, doused with water, slapped, slammed into a wall, given rectal rehydrations without medical need, shackled into painful stress positions and sleep-deprived for about a week by being forced to stand with his hands chained above his head (Savage 2018).

Sen. John McCain, himself a victim of torture in the Vietnam War, did not forget the horrors of his five-and-a-half years-long experience. He attempted to work for change, and offered this piece of advice: "Your last resistance, the one that sticks, the one that makes the victim superior to the torturer, is the belief that were the positions reversed you wouldn't treat them as they have treated you" (Welch 2018). It is not incidental to ask ourselves what we would want if the tables were turned. That needs to be given more credence by those planning or monitoring interrogations.

McCain pointed out the futility of torture in terms of a credible method of extracting information. "In truth," he argues, "most of the C.I.A.'s claims that abusive interrogations of detainees had produced vital leads to help locate Bin Laden were exaggerated, misleading, and in some cases, complete bullshit" (Welch 2018). Assuming the truth of this claim—and we are justified in considering it true—why is it so hard to dislodge the view that torture works? And why is it so hard to take this advice from McCain to heart?

We need to give careful thought to all those members of the health profession who lose sight of their fiduciary duties. They have much to learn from those who hold dear the command "Do no harm" and do not align themselves with the abuse of others, regardless of the goal intended. Caring for their patients should always take precedence over the values of patriotism and combat.

In this book we look at torture as a global issue. Countries the world over treat detainees and prisoners in brutal and degrading ways. Examples abound as to the seriousness of the problem. That doctors have been implicated as participants or enablers in torture raises all sorts of concerns for the profession and for individual doctors.

In the three parts of this book we get an overview of the practices and the concerns that arise with the different ways physical, mental, and psychological torture take place. Global examples give us a sense of the extent of the problem. As a result, we must do some ethical assessment to get a handle on beneficial guidelines to help us find a way out of the circles of hell into the light of day. Given the guilt and shame now coming to the surface, there is reason to hope that lessons can be learned and change can take place.

1.1 Structure of This Book

There are three parts to *Doctors and Torture.* They are;

Part I: Challenges of the Profession
Part II: Boundaries of Torture
Part III: Ethical Assessment

The first part gives an overview of current realities and issues regarding torture and the involvement of doctors. The second part offers a more detailed look at three kinds of torture- physical torture (pain), psychological and mental torture, and isolation. The third part focuses on ethical theories and professional codes that help clarify the duties and responsibilities of medical personnel.

Part I is "Challenges for the Profession." Three chapters set out the challenges. The first, "Torture and Terrorism Overview," opens with the concept of terrorism. The seemingly endless war on terror has functioned as a rationale for all sorts of abusive practices. Drawing from the theoretical insights of Jean Baudrillard, Claudia

1 Introduction

Card, Nel Noddings, and Hannah Arendt, we look at the connection between terrorism and torture and how the former has been used to justify the latter.

Baudrillard discusses popular conceptions and misconceptions around terrorism, such as calling terrorists "cowards." He examines how the dialogue has become polarized and draws our attention to some of the consequences.

We also consider some tools for framing the discussion. Claudia Card offers a theoretical framework for assessing harms and culpability utilizing Deontological and Teleological theories. Both Noddings and Arendt point out that we should not assume evil acts are just committed by "monsters". Rather, quite ordinary people—and quite ordinary doctors—can behave in cruel and abusive ways, assisting torture. All too easily, as we have seen, guards, soldiers, interrogators and, yes, physicians get swept up and lose their moral bearings.

In Chap. 3 we look at global considerations, such as the extent to which torture takes place around the world and the fact that doctors have enabled this practice. A survey of the international realities shows torture is widespread. Countries practicing torture range over the map and though few defend it, many put it to use.

Torture runs the gamut, as we see with the various examples. Doctors have unfortunately played a role—both actively and passively—in allowing torture to become normalized and systemic. The actions of Dr. Shakil Afridi (who used a ruse involving a fake hepatitis vaccine to help the quest for Osama bin Laden) and Drs. Jessen and Mitchell (psychologist-architects of "enhanced interrogation" including waterboarding) highlight the seriousness of the matter.

Drs. Mitchell and Jessen put the SERE (Survival Evasion Resistant Escape) techniques to work, outlining recommended practices to use on suspects, prisoners, and detainees. One issue this raises is to what degree should doctors be complicit, as when they apply their skills when they might lessen the pain of brutal treatment being used. One of the subsequent global issues is what should medical organizations do when doctors cross an ethical boundary and enable torture; e.g., by falsifying records. What sort of response and sanctions are appropriate?

There are two other global considerations examined in this chapter; they are secrecy and rendition. The widespread secrecy and lack of transparency has insured that torture victims are a vulnerable population, thus creating major concerns regarding human rights. Another concern of global significance is the collusion of countries in the practice of rendition, whereby terrorist suspects are sent to countries known to use torture in interrogations. As the chapter demonstrates, these major global issues merit attention to the individual participants as well as organizations and nations.

Chapter 4 focuses on global torture. To grasp the range of abusive tactics found in countries around the world and as recent as 2017, we look at six cases. Each is a picture of brutality; together they testify to a moral failure in terms of human rights. We start with the Mau Mau's fight in Kenya for independence from Britain in the 1950s. We then turn to torture in a Nazi camp in Chile in the 1970s, where human experimentation was practiced as well. Example number three takes us to torture in the Philippines during the Marcos regime and more recently in 2014–2015. In the latter, torture was practiced as a form of entertainment, with human victims as pawns in the game.

Our fourth example is torture in Uzbekistan. There we find the macabre practice of boiling arms or other body parts of victims. Such horrific treatment according to the ex-British ambassador to Uzbekistan is relatively commonplace. Example number five takes us to recent events in Syria, where victims numbered in the thousands. Photographic evidence testifies to the extent of the harms—and deaths. Our last example, number six, is Abu Ghraib prison in Iraq. As with the "Wheel of Torture" game in the Philippines, Americans abused detainees at Abu Ghraib for their own entertainment. It is noteworthy that guilt, shame, and regret are starting to bubble to the surface, making it clear that the victims aren't the only ones harmed by torture.

In Part II we look at the boundaries of torture and the ethical issues they bring to light. Chapter 5 uses Nel Noddings' notion of the three components of evil as a framework of analysis. Each one is applied to torture, recognizing that its aspects of physical pain and mental and psychological suffering are set out by the UN Convention Against Torture in 1984.

We start with diverse examples of the infliction of pain in an attempt to extract information. This includes the "five techniques" practiced on IRA prisoners; namely, hooding, sleep deprivation, white noise, sleep deprivation, and stress positions. A sixth deemed to be a clencher was mock executions and threats to be thrown out of a helicopter.

One change over the years is the increasing use of abusive practices which leave no scars or other physical evidence. Examples are sleep deprivation, isolation, threats to family members, and forced standing. For perpetrators, such tactics offer significant advantages.

In looking at the infliction of pain, one concern is how things can escalate and lead to further abuse. This is "force drift," a concern we look at here. In conjunction, we examine two dimensions of pain, physical and mental (e.g., a "mind virus," as noted in the US Senate Report on Torture). To complicate the matter, clustering adds to the potential harm by combining two or more techniques. This ratchets up the suffering on the part of the victim.

A central concern is the role language has played in allowing inhumane tactics to be authorized. Replacing the term "prisoner" with "detainee" stripped away a host of protections and left "detainees" vulnerable to all sorts of mistreatment, such as indefinite detention.

Richard Jackson offers valuable insights into the way language shapes public consciousness; for example with narratives. It is much easier to abuse our enemies if we do not see them as fellow human beings. As Jackson observes, language and morally defining narratives have made torture more palatable. The impact of language, we should also note, is heightened if medical personnel are viewed as *combatants*. In such an interpretation, "Do no harm" would no longer function as a binding principle.

Chapter 6, the second chapter in Part II, focuses on the helplessness of the victims in the ways they are subjected to degradation and dehumanization. As we see, torture is not just about inflicting pain; it is also about asserting power. Treating others in demeaning ways is an effective vehicle for accomplishing that goal. As the examples testify, the injury can be long-lasting.

1 Introduction

This can be seen in five practices that stand out on the degradation and dehumanization scale. Together they paint a picture of human cruelty, often leaving the victims damaged, and defeated. These are hooding, forced nudity, force-feeding, waterboarding, and humiliation. We will look at each one, some of which are combined, clustered, to magnify their effect.

Hooding and forced nudity both rob detainees of their identity, making maltreatment much easier than if we could see their faces and interact with them. By putting a bag or two over their heads, we shield ourselves from their suffering. And taking away their clothes is an attack on their dignity, thus causing disorientation and instilling fear. Both tactics diminish their sense of self. Furthermore, as Major General George Fay observed, forced nudity can contribute to an increase in sexual assaults. In addition, the use of nude photography adds to the indignity and risk of further abuse.

Force-feeding is also degrading and dehumanizing. It is generally a response to a hunger strike. That it goes against the right of self-determination is examined here. Force-feeding violates the right of competent adults to refuse or withdraw from treatment. So, too, does "rectal hydration," a bizarre form of force-feeding. Doctors and nurses are involved in force-feeding detainees—a controversy that the profession has responded to.

The fourth practice that underscores the helplessness of the victim is waterboarding, alias "simulated drowning." We get a sense of its long history and the role it has played in the interrogation of high-profile suspects. That it is terrifying is born out by the testimony of four Americans who were waterboarded as part of SERE training. Furthermore, that it is said to be a horrific experience did not stop it from being used on high-profile detainees.

The fifth practice we examine is the use of humiliation. Three techniques are especially debilitating—forced shaving (for disorientation and loss of self control), prolonged diapering (an "ego-down" experience as well as physically disgusting), and sexual assault (which is obviously a demonstration of power and cruelty).

In Chap. 7, the last chapter of Part II, we turn to what many consider the worst form of torture, solitary confinement. A disaster in terms of a method of rehabilitation (as intended by the Quakers who put it to use years ago), solitary confinement has brought about any number of harms and no apparent benefits—other than as a vehicle of isolating the individual from the general population.

The days of the Quakers thinking solitary confinement purifies the soul are long gone. In its place is we find a practice that causes both mental and physical suffering having long-term damage. We will look at five cases to get a sense of the potential harm of solitary. They include a lawyer, a detainee, two hostages, three prisoners, and three men on death row. The picture they present is truly alarming.

Each of the five cases indicate why it should be banned or restricted to a matter of hours if absolutely necessary. We then look at the examples of countries that have made changes and, within the U.S., such as the actions of Rick Raemisch, the Colorado governor who put an end to long term solitary in his state.

The last part of the book, Part III focuses on ethical assessment. In Chap. 8 we look at the major ethical theories and how they can help provide a framework for assessing doctors' involvement in torture. This survey and application includes Teleological Ethics (which prioritizes end goals and consequences), Deontological Ethics (which prioritizes moral duty and obligations), Virtue Ethics (which prioritizes a life of virtue and the development of moral character), and Feminist Ethics (which prioritizes relationships and moral agency).

After a brief overview, each theory is applied to the use of torture and the question of doctors' participation. Applying these theories gives us a handle on whether abuse and torture can ever be justified on moral grounds. In order to arrive at an answer, we need to take into account both short-term and long-term consequences.

Chapter 9, the last chapter of the book, focuses on two ethical frameworks and key professional and international guidelines prohibiting torture. We start with two of the major figures in Bioethics, Tom L. Beauchamp and James F. Childress. Their four principles are of central importance in Western Biomedical Ethics. They are: Autonomy, Beneficence, Non-Maleficence, and Justice. Each one is discussed and applied to doctors and torture.

The second ethical framework considered is that of Ethicist Bernard Gert and the ten duties he considers universally applicable. All are then applied to the question of doctors' involvement in torture-showing that doing so would be morally impermissible.

We then look at professional organizations and international codes; all of which prohibit torture. These include the Hippocratic Oath, World Medical Association and its Declaration of Tokyo, the American Medical Association, the International Council of Nurses, the Istanbul Protocol, and Physicians for Human Rights. These organizations and their guidelines emphasize the responsibility and accountability of doctors taking the right path-one to "Do no harm."

Works Cited

Rabbani, Ahmed. 2018, July 26. I'm Stuck in Guantanamo. The World Has Forgotten Me. *Los Angeles Times*. http://www.latimes.com/opinion/op-ed/la-oe-rabbani-guantanamo-prison-torture-20180726-story.html. Retrieved 28 July 2018.

Savage, Charlie. 2018, May 8. 9/11 Planner, Tortured by C.I.A., Asks to Tell Senators About Gina Haspel. *The New York Times*. https://www.nytimes.com/2018/05/08/us/politics/khalid-shaikh-mohammed-gina-haspel.html. Retrieved 10 June 2018.

Welch, Matt. 2018, August 27. What John McCain Taught Us About Torture. *The New York Times*.. https://www.nytimes.com/2018/08/27/opinion/john-mccain-torture-.html. Retrieved 28 August 2018.

Part I
Challenges for the Profession

Chapter 2
Torture and Terrorism Overview

> *Sometimes you have to work with the devil.*
> —Michael Scheuer, CIA, Osama bin Laden unit, noted by Stephen Grey.

Overview

The first part gives an overview of current realities and issues regarding torture and the involvement of doctors, giving us a sense of the challenges facing the medical profession. This chapter opens with the concept of terrorism. The seemingly endless war on terror has functioned as a rationale for all sorts of abusive practices. Drawing from the insights of Jean Baudrillard, Claudia Card, Nel Noddings, and Hannah Arendt, we look at the connection between terrorism and torture and how the former has been used to justify the latter.

We also consider some tools for framing the discussion. Claudia Card offers a theoretical framework for assessing harms and culpability utilizing Deontological and Teleological theories. Both Noddings and Arendt point out that we should not assume evil is just found in "monsters". Rather, quite ordinary people—and quite ordinary doctors—can behave in cruel and abusive ways, assisting torture. All too easily, as we have seen, guards, soldiers, interrogators and, yes, health professionals get swept up and lose their moral bearings.

2.1 Introduction

In his book, *The Spirit of Terrorism*, Jean Baudrillard compares terrorism to a virus. "Terrorism, like viruses, is everywhere" he says, noting that there is a global profusion of terrorism (2003, 10). He doesn't seem far off in that estimate when looking at all the conflicts and violence attributed to terrorism.

Torture is not merely a response to terrorism. However, terrorism is often cited as the motivation for applying interrogation methods that display cruel and degrading treatment—and torture. Although terrorism is not new, after 9/11 the "war on terror" has been at center stage, with no end in sight.

Philosopher Louis P. Pojman relays the scope: "From 1970 to 1995 64,319 terrorist incidents were recorded, half of them attributed to religious extremists. If one excludes state–sponsored terrorism, the twentieth century will be hard to equal in terms of terrorist atrocities" (2003, 135). These numbers are nothing short of staggering. So much unrest is hard to fathom.

Furthermore, given the number of such incidents, it is not surprising that terrorism is viewed as justification for torture. Of course, acts of violence may or may not be associated with terrorism. That label comes into play when there appears to be a political motivation for an act of violence. In that case, it is usually associated with a particular cause, group, cult, or ideological association.

Philosopher Tomis Kapitan sees three components to terrorism; namely, (1) intentional violence or threat of violence, (2) civilian targets, and (3) political motives. To Kapitan's list of components we should add religious or ideological motives, given they can be as powerful as those that are political (see, for example Jessica Stern, *Terror in the Name of God*). As Stern points out,

> When religious terrorist groups form, ideology and altruism play significant roles. Commitment to the goals of the organization, and the spiritual benefits of contributing to a "good cause" are sufficient incentives for many operatives … Terrorism becomes a career as much as a passion (6).

2.2 What is Considered "Terrorism"?

Kapitan sums it up as follows: "Terrorism is the deliberate use of violence, or the threat of such, directed upon civilians in order to achieve political [or ideological] objectives" (2003, 48). A key element in terrorism seems to be the targeting of civilians—"innocents"—as, for example with the bombing of churches or mosques, driving trucks into crowded tourist sites, or releasing a deadly toxin in a subway.

Linguist and political activist, Noam Chomsky modifies this characterization by pointing to the official US definition, which sees terrorism as the calculated use of violence or threat of violence to attain goals that are political, religious, or ideological in nature through intimidation, coercion, or instilling fear (2003, 69). He does not mention targeting civilians; thus bombing Navy ships and military facilities could be seen as an act of terrorism.

Because terrorism tends to be perceived as an evil deed, the responding "war on terrorism" should not be a surprising reaction. Philosopher Claudia Card argues that a "terrorist" is not an identity: "To identify someone as a terrorist is to render a judgment on them, not simply to make a discovery. Not all terrorists have common goals, belong to a unified organization, or have the same opponents," she argues (2003, 178). For Card, the phrase "war on terrorism" should not be viewed in metaphorical terms like the "war on crime." Rather, the "war" in the "war on terrorism" bestows an air of legitimacy that terrorism itself does not have (2003, 175).

Tomis Kapitan points out that groups described as "terrorists" lack legitimacy and terrorist-rhetoric erases any incentive we might have to understand the terrorists' point of view or to examine policies that might have contributed to their grievances. This repudiates any necessity to negotiate, he observes (2003, 53). The result is a divisive *Us versus Them*, with neither side obliged to grant leniency or concede to any demands.

2.3 The Power of Terrorist Rhetoric

Kapitan goes on to argue that terrorist rhetoric increases terrorism and does so in four distinct ways: first, it magnifies the effect of terrorist action by heightening fear; second, those who succumb to the rhetoric contribute to a cycle of revenge and retaliation by endorsing terrorist action on the part of their own government; third, a violent response can have a polarizing effect; and, fourth, it encourages actions that will generate further violence (Kapitan 2003, 53).

In brief, terrorist rhetoric can have an escalating effect in conflicts and political disagreements and therefore we should be careful in its use. Consequently, toning down the language will be more productive than heating things up.

2.4 Tools for Framing the Discussion

As far as torture is concerned, the World Medical Association's Declaration of Tokyo (2016) puts it succinctly:

> For the purpose of this Declaration, torture is defined as the deliberate, systematic or wanton infliction of physical or mental suffering by one or more persons acting alone or on the orders of any authority, to force another person to yield information, to make a confession, or for any other reason.

With that definition in mind, let us see how ethical theories offer some guidance. Claudia Card offers a theoretical framework for assessing harms and culpability utilizing Deontological and Teleological theories. Both Nel Noddings and Hannah Arendt assert that we should not assume evil is just found in "monsters". Rather, quite ordinary people—and quite ordinary doctors—behave in cruel and abusive ways, assisting torture.

It helps to bring in philosophical tools when looking at this issue. Ethical theories in particular can help frame the issue of terrorism and provide insight into how we should proceed. According to Claudia Card, "Insofar as terrorism does its victims intolerable harm and (borrowing from Immanuel Kant's language) treats them merely as means to ends they cannot rationally share, it is an evil" (2003, 171). She contends that terrorism represents a challenge; namely, how to respond effectively without doing further evil. That is a significant concern since responding to terrorism by resorting to torture or "enhanced interrogation" unleashes its own set of evils, as we will see in this book.

Baudrillard asserts that terrorism is immoral and, therefore, it helps to approach it with an understanding of Evil. In his view, "Good is not contrary to Evil, nor does the reverse happen: they are at once both irreducible to each other and inextricably interrelated" (2003, 13). He says that,

> Evil takes hold of [our] unconscious, and realizes by violence what was merely a fantasy and a dream thought! It all comes from the fact that the Other, like Evil, is unimaginable. It all comes from the impossibility of conceiving of the Other— friend or enemy—in its radical otherness, in its irreconcilable foreignness (2003, 62).

One of the ways this form of evil gets expressed is in the separation of victims from perpetrators. Baudrillard raises the question of what is permitted on the part of the victim in responding to an act of terrorism. Does the fact that I have been abused allow me to retaliate with violence? Can I strike back at the perpetrators themselves? What is a morally permissible response on the part of the victims? This is a vital concern, as it lays the groundwork for how suspected terrorists will be treated.

2.5 How Do We Set Boundaries?

Is being a victim of terrorism, as Baudrillard asks, "the perfect alibi" resolving all guilt on the part of the victims? Does it allow them "to use misfortune as though it were, so to speak, a credit card" (61)? As we have seen in the treatment of detainees and prisoners suspected of terrorism, things can get ugly pretty fast. They can become the targets of rage and retaliation.

Baudrillard discusses popular conceptions and misconceptions around terrorism and how the dialogue has become polarized and draws our attention to some of the consequences. One of our failures, he argues, is our inability to regard the Other, the terrorist, as a full-fledged adversary, thereby allowing us to "exterminate him, and obliterated him unceremoniously" (2003, 66).

The use of the word "cowardly" illustrates that mentality in his view. It fails to recognize that there may be something more important than life itself for the terrorist; such as a "destiny, a cause, a form of pride or of sacrifice" (2003, 68). In other words, suicide "martyrs" may warrant more attention than they've been given and not just be dismissed summarily.

Examples of labeling terrorists "cowards" abound. Note the Uzbek truck driver "enemy combatant" terrorist said to commit a "cowardly act of terror" when he drove his vehicle onto a busy New York City bike path in November 2017, killing eight people (Long and Pearson 2017). Note also the "first class coward of Norway," Anders Behring Breivik, who massacred 77 people in the summer of 2011, mostly teenagers (Neba-Fuh 2012).

Moreover, American presidents regularly use this term to apply to terrorists. President Reagan said the 1983 terrorist bombing of the US Embassy in Beirut was "cowardly," President Clinton did the same with the 1998 bombing of the US Embassy in Nairobi, and President George W. Bush called the 9/11 attacks "cowardly" as well (*Slate* 2001). And President Trump condemned the "cowardly terrorist attack" on a mosque in Egypt (Lacy 2017).

2.6 It Pays to Consider the Labels We Use

We might consider the effects of such a dismissive label as "cowards" when assessing terrorists. Presumably the intent is to deprive the actor of satisfaction in his or her deeds, given that few desire to be a coward and therefore would seek to avoid being seen as such. But like Baudrillard indicates, we may be missing the sense of power that comes from intending harm or seeking revenge.

As *Slate* points out, terrorism is inhumane and unforgivable, as well as an offense to morality in international law. What makes for cowards? *Slate* draws a distinction between moral cowardice and physical cowardice, with moral cowards afraid to face the burdens of living and, thus, committing suicide is a cowardly act. "Somehow it isn't enough to abhor terrorism or even to promise to make the terrorist pay dearly. The rules demand that the terrorist be branded a sissy ... the public seems to demand that our presidents call terrorists cowards" (*Slate* 2001).

Claudia Card warns us that "Evil," like "terrorism," is an emotionally laden term and is "liable to abuse in the interest of political manipulation" (Card 2003, 171). Indeed. Any number of abuses are born of a response to the evil of terrorism. For this reason, it pays to examine the concept of evil and what is justifiable in addressing its harms. Card threads together two philosophical frameworks in claiming that "evils are reasonably foreseeable intolerable harms produced by culpable wrongdoing" (2003, 172).

2.7 Grappling With the Concept of Evil

Evils, Card asserts, have two basic elements, harm and wrongdoing, neither reducible to the other. In her view "culpable wrongdoing" includes not just overt acts but also such omissions as negligence, carelessness, recklessness, the failure to attend. This approach unites the Utilitarian concern with maximizing benefits and minimizing harms for the greatest number of people with the Deontological emphasis on individual moral duty and accountability.

Card recommends we find a moral merger between Utilitarians and Kantians (Deontologists). She suggests a combination of the Deontological emphasis on culpability and moral agency with the Utilitarian focus on consequences and the harms incurred—actual damage. This approach offers a way to take into account moral obligations and the consequences of our acts. When it comes to analyzing torture, Card's moral merger may be particularly useful.

Hannah Arendt offers the insight that evil requires us to take action and be vigilant in our moral reasoning and behavior. She points out that quite ordinary-seeming people can commit horrendous deeds. This is seen in her discussion of the trial of Adolf Eichman—a key figure in coordinating the "final solution" (execution/genocide of the Jews in World War II).

In Arendt's view, "There exist many things considerably worse than death, and the S.S. saw to it that none of them was ever very far from their victims' minds and imagination" (1963, 12). She cites the example of the Dutch Jews who in 1941 attacked a German security police detachment in Amsterdam. "Four hundred and thirty Jews were arrested in reprisal and they were literally tortured to death."(1963, 12).

Evil is not just the product of one individual's despicable behavior; we need to cast a wider net. Moreover, we should keep in mind that not all fall under the sway of evil deeds. As Arendt observes, "Politically speaking, it is that under conditions of terror most people will comply but some people will not." (1963, 23). George Kateb (2007) notes that, for Arendt,

> If the mass of people had cared only about the most elementary justice, these policies, which explode all categories of immorality, could not have been carried out. [And], unless morality is a commitment, it can degenerate into mores, merely social behavior. It can, thus, turn into the path of least resistance, which is one of the greatest sources of the strength of recurrent evil.

We might also link this to Kant's notion of the weakness of the will and how easy it is to let others take the reins instead of assuming responsibility for our own decision-making. The onus is on all of our shoulders and not just those of a hero who steps forward to take on the villain. This means that, "resistance to evil includes the practice of genuine morality, and even more, the exertions of vigilant citizens to prevent it" (Kateb 2007).

Philosophers have given little attention to the concept of evil, leaving it to theologians and religious scholars to tangle with. However, Claudia Card and Hannah Arendt have done a service in seeing how evil gets factored into the lives of ordinary people. Both of them point out that we ought not look for savages when trying to spot the manifestation of evil thoughts and deeds. Like the innocuous looking man who beats his wife and children over trivial transgressions, so too "normal," ordinary-seeming people can be pulled into abusive behavior—behavior which all too readily escalates.

2.8 Evil and Torture

Turning to Ethicist Nel Noddings, we can see how we can link the concept of evil to the practice of torture. Nodding's approach offers a way to understand the various methods of interrogation—and torture.

Noddings sees three components to the concept of evil. These are pain, separation, and helplessness, with pain being the most basic form (1989, 95 and 118). The idea that evil is like something out of a horror movie is a bit of a misconception she implies. Her view is in line with that of Hannah Arendt in contending that evil takes shape in the ordinary, rather than the monstrous realm. Specifically,

In describing pain, separation, and helplessness as the trinity of the elemental evils, we feel some of the excitement conveyed by stories of devils ... [However] Evil does not have the stomach-turning stench. Nor does it signal its presence with palpable cold and darkness. We do not fall into it haplessly. Nor does it entrap (possess) us. Rather, we often act willfully and complicity with it (Noddings 1989, 118).

Self-reflection is often lacking in most of us. It's easier to label the actions of others as evil than to judge ourselves and acknowledge our own shortcomings. As Noddings says, "We do not easily look for evil within ourselves" (1989, 194). Nor are we innocent victims of evil curses, insofar as we may willingly participate in morally permissible acts that we later consider inherently vicious or cruel. Acts of torture come to mind, though there are other forms of evil that we might point to. These acts may be on either the individual or societal level. As a result, it pays not to restrict our discussion or proceed with too narrow a focus.

In her book *Women and Evil*, Noddings offers an example on the level of the state—a decision made in World War II by British Prime Minister Winston Churchill to bomb population centers in Germany rather than attack the military. The American bombing of Hiroshima and Nagasaki similarly targeted innocent civilians. The assumption that civilians could become collateral damage did not undergo a public airing—or at least not to the degree Noddings thinks necessary. She views the decision to bomb German cities with a critical eye, particularly given that the escalation of warfare to include civilian targets can have long-lasting repercussions in terms of others following suit.

2.9 Different Kinds of Warfare

The thing about Noddings's example is that there came a point when World War II was considered *over*. In other words, the war had a beginning and an end. Unfortunately, this is not the case with the ongoing war on terrorism. It lacks the tidiness of a war between nations, with their beginnings and ends. It is, thus, an entirely different situation, given that current conflicts often have shifting participants and are not battles between nations. In the case of terrorism, it may not even be clear who is the enemy— the enemy may be elusive or transitory. This fact should not escape our attention.

Part of the problem is that we lack a set of criteria to determine when this "war" will end. Look, for example, at the case of *Rasul v. Bush*. In her ruling on January 31, 2005, Federal Judge Joyce Hens Green made the unsettling observation that, "the government cannot even articulate at this moment how it will determine when the war on terrorism has ended" (Stout 2005).

Plaintiff Shafiq Rasul's lead counsel Joseph Margulies echoes Green's concern. "Even today," he states, the Administration cannot say when the war will end. In fact, it cannot even say how it will know that the end has come" (Margulies 2006, 138). Given the US involvement in the "war on terror" has gone on for over 17 years, Margulies is right to wonder how we'll know when the end has come. It is surely a vital concern.

2.10 Polarization in the War of Terror

Noddings warns us against holding tight to the view that, "The notion that we must choose sides and that our side (whatever it is) must be right and good while the other side is wrong and evil." She points out that, "We saw this phenomenon again and again in the debates over Vietnam." Basically, Noddings asserts, "We do tend to project evil unto others," expressing a view in line with Jean Baudrillard's (1989, 196).

Ironically enough, we may later come to regret actions done in our name. Subsequently, we may be reluctant to apply the label "evil" to what was done on our behalf. We are, however, not so forgiving of the actions of our "enemies." Because of this, the gloves have come off (to quote ex-Vice President Dick Cheney) in the treatment of "enemy combatants" who have been shown little mercy. Attaching the label "terrorism" to an act of violence by an individual or group opens the door to retaliatory actions that can move us ever closer to torture and abuse.

Whereas pain comes to mind most readily when we think of evil acts, Noddings suggests:

> Separation [isolation] is evil because of the deep psychic pain it causes, and the fear of separation makes human beings vulnerable to all sorts of further evils. Helplessness too is associated with psychic pain, and we must consider its deliberate infliction a great moral evil (1989, 118).

As we will examine Part II, we have seen the infliction of pain, isolation (= separation) and forced helplessness in the treatment of prisoners and detainees. These three types of evil acts have been tolerated if not justified due to the fear of terrorism.

One manifestation of this fear, suggests Baudrillard, is an obsession with security, which has its own set of limitations. He argues that all the security strategies are merely extensions of terror, that, basically, the fear of terrorism has unleashed its own kind of terrorism. In his view, "the real victory of terrorism is that it has plunged the whole of the West into the obsession with security" into a veiled form of perpetual terror.

In other words, "The specter of terrorism is forcing the West to terrorize itself—the planetary police network being the equivalent of the tension of the universal Cold War, of the fourth world war imprinting itself upon bodies and mores" (2013, 81–82). As a result, Baudillard holds, the consequences of the widespread fear of terrorism are sobering.

2.11 Problems With Polarization in the War on Terror

There is an inherent pressure to accept actions we may later consider ill-chosen and morally repulsive. This pressure comes from an identification of *Us versus Them, Right versus Wrong, Good versus Evil*. This pressure is not to be underestimated, as we will see when we turn our focus to the medical personnel who enable cruel and degrading actions.

Philosopher Paul Ricoeur thought it inevitable that there would be moral tangles around evil. He believes people are fundamentally fallible, so we ought to abandon the ideal of total certainty (Reagan, 9). One factor in making torture more palatable is in the dehumanization of detainees. This is in line with Hannah Arendt's assessment of the slide into Evil.

Detainees' low moral status was sealed when the term "prisoners" was replaced with "detainees" (Teays 2008, 2014). For example, the Navy's Human Research Protection Program report defines "prisoner" as, "Any individual (other than Captured or Detained Personnel) involuntarily confined or detained in a penal institution" (2006, 4).

If we shine the moral spotlight on the issue, several things are illuminated. One is that a great deal of time and energy has been expended in manipulating language to serve our ends. Instead of terms like "prisoner" we are given "detainees," a term with much less moral and legal significance. The division between "Them and Us" was absolutely necessary in order or the term "detainee" to stick.

This and other linguistic manipulation such as calling detainees *unlawful* or *unprivileged* combatants, laid the groundwork for abusive policies to be put in place—and justified. Attention could then turn to deciding what is acceptable treatment. Here's where severing the moral ties to "prisoners" was a gold mine. We have had vastly more detainees than prisoners of war, and their resulting diminished moral status comes with that. Their indefinite detention and mistreatment indicate the lack of attention to human rights concerns.

Claims of abuse raise serious issues regarding human rights. Consider the case of Majid Khan. Citing the April 2019 pre-sentencing hearings on the treatment of Khan at Guantanamo Bay, Carol Rosenberg of *The New York Times* reports, "Khan [spoke of what] the C.I.A. did to him. By his account, he was beaten, hung naked from a wooden beam for three days with no food, kept for months in darkness, and submerged, shackled and hooded, into a tub of ice and water" (2019). How much testimony is allowed in his trial remains to be seen, but such claims of cruel treatment need to be taken seriously.

Even the few prisoners of war may fail to get proper care—the care spelled out by international codes. We see this, for instance, with the "care" prisoner of war John Walker Lindh received. *The New York Times* (2002) reports that a medic checked Lindh, who had a bullet in one of his legs and was malnourished, but his requests for medical attention to his wounds and for additional food were refused. Tied up with duct tape and deprived of clothing, Lindh was initially confined to a metal container in frigid weather conditions. He was also threatened with death and torture (*The New York Times* 2002).

The ill-treatment of John Walker Lindh is but one of many examples of medical personnel falling short of their fiduciary duties to their patients in the war on terror. Some, like Lindh's medic, showed neglect; he didn't do what one would expect of a doctor or caregiver. Others have taken a more active role in assisting interrogations or enabling torture.

Jeff Evans (2005) reports that,

> At Guantanamo Bay and Abu Ghraib prison, mental health professionals, such as psychiatrists and psychologists, are known to have observed interrogations, provided interrogators with the medical records of detainees, and in some cases, developed individualized interrogation plans or provided advice on how best to conduct an interrogation.

2.12 How Do We Bolster the "Ethics" in Medical Ethics?

Evans notes how attempts were made to dilute the "Ethics" in Medical Ethics. He cites Bioethicist M. Gregg Bloche's observation that the civilian leadership at the Pentagon argued that when physicians and other health professionals participate in the interrogation process and other non-therapeutic roles, medical ethics do not apply. Since they are not then acting as physicians or health professionals, medical codes are inapplicable. However, "This is a deeply disturbing argument with little or no precedent elsewhere," Bloche said (Evans 2005).

Indeed. The argument that medical codes are fluid in the sense of being context-dependent is both specious and dangerous. Whether you are a doctor in a detention facility instead of a hospital you are still bound by professional ethics. Furthermore, doctors present at interrogations may "merely" be on the sidelines. But at any time they may have to assume a medical role; for example, if a detainee has an adverse reaction to their treatment, suffers injuries, or otherwise needs medical care.

The notion that doctors or psychologists who are passive observers of abusive interrogations have no obligation to intervene and stop or report abuse is an affront to deeply-held ethical theories—such as Deontological ethics—and basic human rights. In addition, the ethical codes of their professions have set out clear standards of conduct. James H. Bray, president of the American Psychological Association, asserts that, "Psychologists have an obligation to intervene to stop torture or abuse, and a further obligation to report any instance of torture or abuse" (American Psychological Association 2009).

2.13 The Path to Torture

We should not underestimate the attempts to avoid having an action judged as "torture" or as evil. We need only look at the so-called "torture memos" of the Bush administration to see evasive maneuvers regarding what is or is not permissible. As Humanities Professor Colin Dayan notes in discussing the sleight-of-hand use of language,

> The writers of the [March 6, 2003] memorandum [sent by Assistant Attorney General Jay S. Bybee to White House counsel, Alberto Gonzales] justified their narrowing of the [Geneva] Conventions' protections by arguing that "the meaning of the term "degrading treatment" was "vague and ambiguous" (2007, 65–66).

That's an interesting argument. *To rephrase*: assuming the term "degrading treatment" is "vague and ambiguous," we are then "justified" in shrinking the Geneva Conventions' protections against abuse (degrading treatment).

Does that conclusion necessarily follow? If what constitutes "degrading treatment" is either vague or ambiguous, why wouldn't it follow that the Geneva Conventions' protections are actually broader than has been understood to be the case? The torture memos bear more resemblance to language games in *Alice in Wonderland* than reasoned, defensible argument.

US State Department's legal adviser, John B. Berlinger III, echoes this concern. He complained that the language in the term "cruel, inhumane or degrading treatment" was so ambiguous that it was "difficult to state with certainty and precision what treatment or punishment would be prohibited or permitted" (Dayan 2007, 81).

Dayan defines torture as any "act committed by a person acting under the color of law specifically intended to inflict severe physical or mental pain" (2007, 65–66). This definition has been stretched and twisted so that different forms of abuse might escape being labeled "torture." Once that is accomplished, what is allowable can expand. The range of such cases will be examined in the next part, when we look at torture and its variations.

2.14 The Terms We Use Make All the Difference

Don't underestimate the importance of the terms of the discourse given the potential consequences of the meaning of the concepts we use. Social Anthropologist Tobias Kelly discusses the language of torture and the difficulties nailing down the concept. "The recognition of torture presents unique challenges," he argues.

> Torture's particular stigma, as one of the most universally recognized violations of human rights raises the stakes for those states accused of torture. Very few, if any, states willingly admit that they participate in torture. Furthermore, despite its apparent moral absolutism, torture remains a notoriously slippery category to define because its meaning constantly shifts under pressure (2009, 778).

Kelly concludes that, "Any attempt to recognize torture must therefore overcome serious political, legal, and epistemological hurdles" (2009, 778). This is especially the case in light of the weight the label "torture" has assumed. And so it is that considerable energy has been expended to avoid torture's black marks.

Given the stakes are high with regard to human rights, it should not be surprising that Jay Bybee should be reluctant to have the term "degrading treatment" applied to what was sanctioned by the U. S. government. The quest to loosen the restrictions around abuse and inhumane treatment without crossing legal boundaries, ethical codes, or international treaties, has led them down a perverse path. They contorted their reasoning, as well as the use of language, to accomplish their ends. And that's not all.

2.15 Doctors Add Legitimacy

Medical personnel were called upon to add credibility. Bybee required doctors monitor enhanced interrogations. This move pulled in doctors so they had a role to play in the war on terror. It also provided a veil of legitimacy to spread the blame should things go sideways. And things surely did go sideways. Let's see how.

Reliable evidence indicates that physicians have assisted torture in at least 70 countries without facing any punishment (Miles 2014). The extent of such widespread participation of doctors is startling and deserves to be examined very closely. Colin Dayan (2007) points out one of the consequences of this role doctors were to assume. This was to provide treatment for the injuries sustained as a result of torture, which had the effect of allowing torture to continue "just as the British Medical Association had warned a generation earlier."

Suggesting an unconscious duplicity, she argues that, "It is not even clear, though, that the medical personnel were aware of how they were being manipulated" (2007, 65–66). Of course, we might wonder: If they weren't aware they were being manipulated, why weren't they? And if they were aware why did they stay silent? What factors came into play?

Robert Jay Lifton (2004) remarks on "doctors' vulnerability to being socialized to abusive environments and to engage in destructive behavior." He recommends that, "We need to learn all that we can about abuses by doctors everywhere" (2004). Further, we need to address potential conflicts that medical personnel face in the war on terrorism.

2.16 Conflicting Loyalties, Messy Dilemmas

Lifton points out that conflicts come into play on a regular basis; namely, "To be a military physician is to be subject to potential moral conflict between commitment to the healing of individual people, on the one hand, and responsibility to the military hierarchy and the command structure, on the other" (2004).

This is the conflict of "dual loyalties" and an ethical dilemma that is not easy to resolve. It is one of the many challenges doctors face when treating detainees qua patients and otherwise participating in situations that test their loyalties and professional behavior.

Don't underestimate the importance of doctors' fiduciary duties to their patients. Violations of those duties should be taken seriously. Note, for example, Bioethicist Michael J. Gross's (2013) examination of force-feeding and the public interest. Gross states:

> It is unimaginable that any decent society today would leave 10 Irish Republican Army hunger strikers to die of starvation as the British did in Northern Ireland in 1981. Accounts of their slow and anguished deaths are harrowing, and no rights-respecting government or medical association should ever permit a repetition of that event. Instead, we should think about how to feed hunger strikers humanely. (104)

Doctors face difficult decisions about how best to weigh competing values, such as patient autonomy, the patient's best interests, and the needs or interests of others. This requires health professionals putting their moral reasoning to task, especially regarding the moral status of their patients.

As Bioethicist Steven H. Miles points out, "Simply put, health professionals are accountable for the health of their patients, regardless of the fact of imprisonment" (2009, 65). It should not be the position of doctors to judge the moral status of their patients—all should receive the same level of treatment.

2.17 Doctors' Susceptibility to Pressure

Another concern is the exploitation of doctors. For example, Jonathan Cook (2009) reports on a prison doctor, who "under pressure from interrogators, agreed to retract a written recommendation that a detainee be immediately hospitalized for treatment." Such cases of skewed priorities merit our concern. Surely it should be a red flag for doctors as well. However, "Most physicians who abet torture simply rationalize their work," Miles asserts (2009, 30).

Doctors play a crucial role; their presence adds credibility to decisions and actions around the treatment of detainees. Anat Litvin of the Physicians for Human Rights puts that under the spotlight: "We believe that doctors are used by torturers as a safety net" (Cook, 2009). The consequences of that reality can have far-reaching harms.

Miles maintains that, "Twentieth-century torture has … [resulted] in the abuse of many innocent or ignorant persons. Societies that torture light the fuse on a real time bomb" (2009, 13). Torture casts a long shadow on those who have been brutalized and creates a substantial public relations problem.

2.18 Torture Casts a Long Shadow

The photo from Abu Ghraib of a detainee being led like a dog on a leash elicited outrage from the Arab world and other photos of barbaric abuse compounded the problem. Some, like the image of the hooded detainee (nicknamed "Gilligan") standing on a box with arms extended apparently about to be electrocuted have become icons of torture. There is likely to be a "real time bomb," as Miles puts it, on that photograph alone. The photographic evidence of global torture demonstrates the difficulty of putting on the brakes—or, as Miles says, stopping the fuse from being lit—once torture is set in motion.

This waffling on whether an abusive act qualifies as "torture" continues, as we will be discussing in this text. A great deal follows from the way we define our terms. Ethicist Tom Koch says, on the one hand, the US and some of its allies are evidently engaged in the systematic and widespread use of torture, but, on the other hand, the US administration is reluctant to defend this publicly. Quite the contrary. It has even gone on record condemning the use of torture in states such as Syria and Egypt to which it has, nevertheless, 'rendered' terror suspects (2006, 131).

2.19 Conclusion

Torture is a bit like lying. Most people say it's wrong, but it may still be used to suit their needs. Once that moral threshold has been crossed it's easier to continue—and escalate. Nevertheless, as Bioethicist Jonathan H. Marks argues, the deployment of aggressive interrogation strategies violates fundamental norms of international human rights law and the laws of war—and the International ban is absolute (2005, 18 and 22). The prohibition of torture separates law from brutality, argues Jeremy Waldron (noted by Marks 2005, 18). Torture is not an effective way to extract information or build rapport. It may function as punishment, but that hardly justifies its use.

Of fundamental importance to this text and to the issue of torture is the concern raised by Tom Koch:

> In theory, "all doctors have obligations to report human rights abuses" ... they have had, at least since Nuremberg, the obligation to refuse orders that would require them to commit abuses. Why, then, were commissioned medical officers not the first to draw military and public attention to torturous acts? Why did they not refuse assignments that violated professional oaths and international covenants? (2006).

As we proceed we will try to answer those questions and see what should be done to provide guidance as doctors navigate this territory.

Works Cited

American Psychological Association. 2009, April 22. Saying It Again: Psychologists May Never Participate in Torture. Press Release. http://www.apa.org/news/press/releases/2009/04/editorial-bray.aspx. Retrieved 10 June 2018.

Arendt, Hannah. 1963. *Eichmann in Jerusalem: A Report on the Banality of Evil.* New York: Penguin.

Baudrillard, Jean. 2003. *The Spirit of Terrorism.* London: Verso.

Card, Claudia. 2003. Making War on Terrorism in Response to 9/11. In *Terrorism and International Justice*, ed. James P. Sterba. Oxford: Oxford University Press.

Chomsky, Noam. 2003. Terror and Just Response. In *Terrorism and International Justice*, ed. James P. Sterba. Oxford: Oxford University Press.

Cook, Jonathan. 2009, June 30. Israeli Doctors Colluding in Torture of Palestinian Detainees. *The Electronic Intifada.* https://electronicintifada.net/content/israeli-doctors-colluding-torture-palestinian-detainees/8321. Retrieved 10 June 2018.

Dayan, Colin. 2007. *The Story of Cruel and Unusual.* Cambridge, MA: The MIT Press.

Evans, Jeff. 2005, September 15. Experts Call for Detainee Interrogation Guidelines. *Family Practice News.* https://www.mdedge.com/familypracticenews/article/26391/mental-health/experts-call-detainee-interrogation-guidelines. Retrieved 10 June 2018.

Works Cited

Gross, Michael J. 2013, July 11. Force-Feeding, Autonomy, and the Public Interest. *New England Journal of Medicine,* 103–105. https://www.nejm.org/doi/10.1056/NEJMp1306325. Retrieved 10 March 2019.

Kapitan, Tomis. 2003. The Terrorism of "Terrorism". In *Terrorism and International Justice,* ed. James P. Sterba. Oxford: Oxford University Press.

Kateb, George. 2007, Fall. Existential Values in Arendt's Treatment of Evil and Morality. *Social Research* 74 (3): 811. https://www.jstor.org/stable/40972127?seq=1#page_scan_tab_contents. Retrieved 10 June 2018.

Kelly, Tobias. 2009, August. The UN Committee Against Torture: Human Rights Monitoring and The Legal Recognition of Cruelty. *Human Rights Quarterly* 31: 778. https://muse.jhu.edu/. Retrieved 30 June 2018.

Koch, Tom. 2006, May. Weaponizing Medicine. *Journal of Medical Ethics* 32 (5): 249–252. https://www.jstor.org/stable/27719620?seq=1#page_scan_tab_contents. Retrieved 10 June 2018.

Lacy, Acela. 2017, November 24. Trump Condemns 'Cowardly Terrorist Attack' on Mosque in Egypt. *Politico.* https://www.politico.eu/article/donald-trump-condemns-terrorist-attack-egypt-mosque/. Retrieved 10 March 2019.

Lifton, Robert Jay. 2004, July 29. Doctors and Torture. *New England Journal of Medicine* 351: 415–416. https://www.nejm.org/doi/full/10.1056/nejmp048065. Retrieved 10 June 2018.

Long, Colleen and Jake Pearson. 2017, November 1. 'Cowardly Act of Terror': Truck Driver Kills 8 on Bike Path. *AP News.* https://apnews.com/aa83dfe6157f4214a5e92efaba4142c9. Retrieved 10 June 2018.

Margulies, Joseph. 2006. *Guantanamo and the abuse of presidential power.* New York: Simon & Schuster.

Marks, Jonathan H. 2005, July–August. Doctors of Interrogation. *The Hastings Center Report* 35 (4): 17–22. http://www.jstor.org/stable/3528822. Retrieved 10 June 2018.

Miles, Steven H. 2009. *Oath Betrayed: America's Torture Doctors.* Berkeley, CA: University Of California Press.

———. 2014, January 22. The New Accountability for Doctors Who Torture. *Health and Human Rights Journal.* https://www.hhrjournal.org/2014/01/The-New-Accountability-For-Doctors-Who-Torture/. Retrieved 10 June 2018.

Neba-Fuh, Stephen. 2012, April 17. Anders Behring Breivik- The Cowardly terrorist of Norway. *Voices of the Oppressed.* http://www.nebafuh.com/2012/04/anders-behring-breivik-the-cowardly-terrorist-of-norway.html. Retrieved 10 June 2018.

The New York Times. 2002, February 2. Excerpt from Lawyers' Filing for Lindh: 'Threatened Him with Death'. *The New York Times.* https://www.nytimes.com/2002/02/06/national/excerpt-from-lawyers-filing-for-lindh-threatened-him-with-death.html. Retrieved 10 June 2018.

Noddings, Nel. 1989. *Women and Evil.* Berkeley, CA: University of California Press.

Pojman, Louis P. 2003. The Moral Response to Terrorism and Cosmopolitanism. In *Terrorism and International Justice,* ed. James P. Sterba. Oxford: Oxford University Press.

Reagan, Charles E. Personal Identity. *Scribd.* https://www.scribd.com/document/333134997/Personal-identity-Charles-Paul-Ricoeur-pdf. Retrieved 10 June 2018.

Rosenberg, Caroline. 2019, April 6. Guantánamo Trials Grapple With How Much Evidence to Allow About Torture. *The New York Times.* https://www.nytimes.com/2019/04/05/us/politics/guantanamo-trials-torture.html?smid=nytcore-ios-share. Retrieved 6 April 2019.

Slate. 2001, September 11. In What Sense Are Terrorists Cowards? *Slate.* http://www.slate.com/articles/news_and_politics/chatterbox/2001/09/in_what_sense_are_terrorists_cowards.html. Retrieved 10 June 2018.

Stern, Jessica. 2003. *Terror in the Name of God: Why Religious Militants Kill.* New York: HarperCollins.

Stout, David. 2005, January 31. U.S. Denies Guantánamo Inmates' Rights, Judge Says. *The New York Times.* https://www.nytimes.com/2005/01/31/politics/us-denies-guantnamo-inmates-rights-judge-says.html. Retrieved 10 June 2018.

Teays, Wanda. 2008. Torture and Public Health. In *International Public Health Policy and Ethics*, ed. Michael Boylan. New York: Springer.

———. 2014. Torturous Deeds: Crossing Moral Boundaries. In *Global Bioethics and Human Rights: Contemporary Issues*, ed. Wanda Teays, John-Stewart Gordon, and Alison Dundes Renteln. Lanham, MD: Rowman & Littlefield.

U.S. Department of the Navy. 2006, November 6. Protection of Human Subjects and Adherence to Ethical Standards in Dod Supported Research. http://www.fas.org/irp/doddir/navy/secnavinst/3900_39d.pdf. Retrieved 10 June 2018.

World Medical Association. 2016, October. Declaration of Tokyo – Guidelines for Physicians Concerning Torture and other Cruel, Inhuman or Degrading Treatment or Punishment in Relation to Detention and Imprisonment. *WMA.org*. https://www.wma.net/policies-post/wma-declaration-of-tokyo-guidelines-for-physicians-concerning-torture-and-other-cruel-inhuman-or-degrading-treatment-or-punishment-in-relation-to-detention-and-imprisonment/. Retrieved 10 March 2019.

Chapter 3
Global Considerations

> *I don't think we had any idea doctors were involved to this extent, and it will shock most physicians.*
> —George J. Annas.

Overview

The first unit gives an overview of current realities and issues regarding torture and the involvement of doctors, giving us a sense of the challenges facing the medical profession. Here in this chapter we look at global considerations, such as the extent to which torture takes place around the world and the fact that doctors have enabled this practice. Countries practicing torture range over the map and though few defend it, many put it to use. And doctors have played a role—both actively and passively—in allowing torture to become normalized and systemic. For example, Drs. Jessen and Mitchell (psychologist-architects of "enhanced interrogation" including waterboarding) highlight the seriousness of the matter.

Drs. Mitchell and Jessen put the SERE (Survival Evasion Resistant Escape) techniques to work, outlining recommended practices to use on suspects, prisoners, and detainees.

One issue this raises is to what degree should doctors be complicit; e.g. to apply their skills when they might lessen the pain of brutal treatment being used—and what should medical organizations do when doctors enable torture; e.g., by falsifying records. What sort of response and sanctions are appropriate? I discuss these concerns and examine two other global considerations; namely, secrecy and rendition. The widespread secrecy and lack of transparency creates human rights concerns. Another concern of global significance is the collusion of countries in the practice of rendition, whereby terrorist suspects are sent to countries known to use torture in interrogations. As the chapter demonstrates, these major global issues merit attention to the individual participants as well as organizations and nations.

3.1 Introduction

At this point it is well known the torture is on a global scale. According to Amnesty International, torture is widespread in more than a third of countries (Lepora and Millum 2011, 38). Wisnewski and Emerick (2009) point out that countries utilizing torture range over the map. The list includes Afghanistan, Algeria, Argentina, Cambodia, Chile, and China, Egypt, England, France, and Germany, Indonesia, Israel, Italy, and Mexico, Nigeria, Saudi Arabia, South Africa, and the Soviet Union, Thailand, Turkey, the United States, Uruguay, and Zimbabwe —and many more.

Even countries like Canada have a less than stellar record. We see this in the complicity in Canadian citizen Maher Arar being arrested at the JFK airport and "rendered" to Syria to face being beaten with 2″ electric cables and confined to a rat infested 3′ × 6′ × 7′ cell. The Canadian government later agreed to a $10.5 million settlement for Arar—falsely accused of terrorism.

Alex J. Bellamy, Professor of Peace and Conflict Studies, observes that torture is a moral anomaly in that, while few defend it, many countries use it. This results in the systematic and widespread use of torture by those unwilling to endorse it publicly (2006, 131).

As we will see in this text, doctors have enabled torture in both passive and active ways. It is now clear, as Bioethicist George J. Annas points out, that doctors play a more significant role than has been recognized (as noted by History Commons 2009). Both the US Army's Report on Detainee Medical Operations and the Major General George Fay's Report reveal that medical personnel were among the 54 personnel found responsible or complicit in the abuse at the Abu Ghraib prison in Iraq (Kiley 2005; Fay 2004).

Throughout this book we will look at a variety of examples of physician involvement in torture and abuse. One of many is the participation in ending hunger strikes. George J. Annas, Sondra S. Crosby, and Leonard H. Glantz deplore such behavior. In their view,

> Hunger strikers are not at tempting to commit suicide. Rather, they are willing to risk death if their demands are not met. Their goal is not to die but to have perceived injustices addressed. The motivation resembles that of a person who finds kidney dialysis intolerable and discontinues it, knowing that he will die ... Physicians who participate in this nonmedical process become weapons for maintaining prison order ...
>
> Using a physician to assault prisoners no more changes the nature of the act than using physicians to "monitor" torture makes torture a medical procedure. Military physicians are no more entitled to betray medical ethics than military lawyers are to betray the Constitution or military chaplains are to betray their religion. (2013, 102)

When health professionals fail to "Do no harm," the consequences are more extensive than damaging reputations. The effects are considerable and, thus, it is imperative that we grasp the significance of this moral failing. And we ought not avert our eyes from the diverse examples that come to light. Here are three:

Example #1 Interrogation of Guantánamo Bay Detainee Mohammed al-Qahtani
Some of the medical involvement in torture defies belief. In one of the few actual logs we have of a high-level interrogation, that of Mohammed [al-Qahtani] ... doctors were

present during the long process of constant sleep deprivation over 55 days, and they induced hypothermia and the use of threatening dogs, among other techniques. According to [Steven H.] Miles, medics had to administer three bags of medical saline to [al-Qahtani] — while he was strapped to a chair — and aggressively treat him for hypothermia in the hospital. They then returned him to his interrogators (Sullivan 2006).

Example #2 Torture and the 2008 Zimbabwe Election

Human Rights Watch documented cases of violence during the 2008 [Zimbabwe] elections, showing that the ZANU-PF-led government was responsible, at the highest levels, for widespread and systematic abuses that led to the killing of up to 200 people, the beating and torture of 5000 more (Human Rights Watch 2011). For the most part health professionals in Zimbabwe were generally silent against the abuses (London et al. 2008).

Example #3 Long-Term Torture of Mentally Ill Patients in Ontario, Canada

Patients at a maximum-security mental-health facility in Ontario were tortured by medical doctors over a 17-year period in unethical and degrading human experiments, a judge has ruled in a lawsuit. The techniques used on the patients between 1966 and 1983 included solitary confinement, as treatment and as punishment; the administration of hallucinogens and delirium-producing drugs, including LSD; and brainwashing methods developed by the CIA, according to Justice Paul Perell of the Ontario Superior Court of Justice (*The Globe and Mail* 2017).

3.2 Doctors' Involvement in Torture

So where do medical personnel come in—how are they implicated? According to Jesper Sonntag (2008),

> Testimonies from both torture survivors and doctors demonstrate that the most common way doctors are involved is in the diagnosis/medical examination of torture survivors/prisoners. And it is common before, during and after torture. Both torture survivors and doctors state that doctors are involved during torture by treatment and direct participation. Doctors also falsify journals, certificates and reports.

Sonntag rightfully observes that doctors' involvement in torture has serious consequences for both survivors and doctors. In addition, the impact extends beyond the individual doctor to the profession as a whole. "When the unambiguous role of the doctor as the protector and helper of people is questioned," Sonntag argues, "it affects the medical profession all over the world" (2008). At that point we have an institutional problem, not just the ethical transgression of individuals.

Bellamy (2006) offers some background information by looking at the case of Algeria in the 1950s. At that time France instituted torture as a response to "an entirely new form of warfare" in Algeria. The view was that traditional fighting methods were no longer effective and, thus, "unconventional methods" were required.

That opened a door of abuse whereby the torture of Algerians was committed more and more frequently. "Although at first using such methods only in exceptional cases," Bellamy explains, "French torture of Algerians was committed on an increasingly regular basis until it became a normal part of interrogation" (2006, 128). In short, torture became normalized.

In 1955 France came under pressure to restrict torture, but only to a point. Permissible methods could include the use of electric shocks and the so-called 'water technique' (holding the victim's head under water until he/she nearly drowns). Some said this was "not quite torture" (Bellamy 2006, 128). We can but wonder where the lines are drawn.

3.3 Exceptions Can Become Normalized

And so it seems that, once torture is allowed in the exceptional case, moral reluctance erodes and it is easier to use it on an everyday basis. As torture then gets applied to the less extreme cases, it gets harder to stop it from becoming systematic.

One way that gets sanctioned is in redefining the terms and restricting the use of the term "torture." As we will see, there are many ways that have been employed to draw exclusions around what counts as torture. Consequently, issues like employing evasion and manipulation of language should come under the spotlight—and be viewed with skepticism.

In reality, torture runs the gamut. It ranges from the use of hypothermia, waterboarding, and electric shock to sleep deprivation, solitary confinement, and shackling prisoners in twisted positions, among a myriad of other abuses. It also includes beating or slapping to "induce surprise, shock, or humiliation", and cutting off a prisoner's healthy ear or limb as punishment, as noted by Physician Chiara Lepora and Bioethicist Joseph Millum (2011, 39).

3.4 And Doctors Get Pulled In

Let us not overlook the ways doctors play a part. One doctor whose actions had grave consequences is Shakil Afridi. He is the Pakistani doctor who assisted the CIA in a fake hepatitis vaccination program in the quest for al Qaeda leader Osama bin Laden. They sought DNA samples to confirm whether bin Laden was hiding out at a particular compound. Regardless of the short-term gains, the long-term consequences were devastating for the eradication of polio as well as the safety of medical workers in the field. The latter were now viewed with suspicion and not to be trusted, resulting in the deaths of some of the workers.

In January, 2012 then U.S. Defense Secretary Leon E. Panetta declared that Afridi was "very helpful" in the CIA's quest and should not face treason charges in Pakistan (Mazzetti 2012). That this was a setback for vaccination efforts did not receive the weight it deserved. "Taliban commanders in two districts banned polio vaccination teams, saying they could not operate until the United States ended its drone strikes," notes Journalist Donald G. McNeil, Jr. (2012). Moreover, as Heidi Larson of *The Guardian* points out,

> Distrust about the polio vaccine and its western providers were rampant in some communities, and suspicions about CIA links with the polio vaccination campaigns, and rumours they were a front for the sterilising of Muslims, had been around for a decade after 9/11.

After years of working to dispel myths about CIA links to the polio eradication efforts ... all of the work seemed fruitless (Larson 2012).

3.5 Psychologists Cross the Ethical Boundary Too

Psychologists have also played a key role in abuse and torture. The most high-profile case involves the two individuals who set the parameters for "enhanced interrogation" and were called the "architects of the interrogation program." These are Drs. John Bruce Jessen and James Mitchell. They insisted that they were "reluctant participants" but, nevertheless, said the practices were effective in dealing with "resistant detainees." (*The New York Times* 2017)

They also profited from their participation: Their involvement netted their business $81 million from the CIA. *The New York Times* (2017) reports that the two psychologists dusted off the SERE [Survival Evasion Resistance Escape] program and put it to use on suspects.

> The techniques were largely adapted from those the psychologists had used to train American soldiers in survival schools to resisting brutal interrogations by hostile forces that were violating the laws of war ...
> Abu Zubaydah [erroneously thought to be a top leader of Al Qaeda] taken into custody in 2002, was the first detainee to be waterboarded. At a secret C.I.A. jail in Thailand ... agency leaders chose to use extreme physical force to break him (2017).

The psychologists were directly involved—hands on, as it were.

> Drs. Mitchell and Jessen were sent to the jail to carry out the techniques, including waterboarding ... [Zubaydah] underwent the procedure 83 times over a period of days; at one point he was completely unresponsive, with bubbles rising from his mouth, according to the Senate [Intelligence Committee] report (*The New York Times* 2017).

Evidently when those at the prison wanted to end the waterboarding sessions of Zubaydah, supervisors ordered them to continue. Jessen and Mitchell said they were under considerable pressure, as if that were a legitimate excuse:

> "They kept telling me every day a nuclear bomb was going to be exploded in the United States and that because I had told them to stop, I had lost my nerve and it was going to be my fault if I didn't continue," Dr. Jessen testified.

> Dr. Mitchell said that the C.I.A. officials told them: "'You guys ... are pussies. There was going to be another attack in America and the blood of dead civilians are going to be on your hands" (*The New York Times* 2017).

Waterboarding and water dousing (which they endorsed the use of) are seen as especially torturous and frightening to undergo. "Dr. Mitchell, once said that most people would prefer to have their legs broken than to be waterboarded" (*The New York Times* 2017). Given that assessment, their approval of waterboarding stretches the imagination.

Waterboarding was not the only type of brutal interrogation technique put to use. There were other horrific abusive practices. Mohamed Ben Soud, for instance, told of being placed in a wooden box that was poked through with holes, slammed against a wall and doused with buckets of ice water while naked and shackled (Fink 2017).

As for the effect of such interrogation techniques: in 2016, *The New York Times* found a pattern of long-term psychological damage among dozens of former detainees subjected to vicious treatment. This counters Mitchell's claim that waterboarding may "suck" but "I don't know that it's painful." He considers it better described as "distressing" (*The New York Times* 2017). Clearly that's an understatement.

3.6 From the Few to the Many

Furthermore, the brutality is systemic. These are not the acts of a few yahoos playing out revenge fantasies in the field. The perpetrators are on a much grander scale, orchestrated from the top and monitored by medical personnel, such as Dr. Afridi and Drs. Jessen and Mitchell. Independent oversight was sorely lacking.

The Red Cross had only limited access to prisons but could, in any case, infer that, "We were dealing here with a broad pattern, not individual acts." Pierre Kraehenbuel, the Red Cross operations director, added, "There was a pattern and a system" (*The Washington Post* 2004). The practice was intentionally devised.

Given the destruction of dozens of interrogation tapes, mostly of the interrogation of Abu Zubaydah, we may never know the extent of the abuse. However, we've seen enough to know that medical caregivers played a part. This is not to say that participating in torture and degrading treatment of suspects was necessarily an easy call for doctors.

3.7 Moral Quandaries in Assisting Torture Victims

It may not always be clear where one fiduciary duty ends and another begins. Journalist Josh Clark (2011) points out that physicians face a very real dilemma of whether to provide medical assistance to people in these circumstances.

> On the one hand, a doctor who provides medical services to those who are being subject to torture are inherently complicit to that torture. In fact, their presence not only condones it, but may allow it, since in extreme cases, a physician's care could keep a tortured prisoner alive to face more torture.
>
> On the other hand, a doctor who refuses to provide medical care for a tortured person is shirking the other foundational tenet of medicine, to provide aid to anyone who needs or wants it (Clark 2011).

Lepora and Millum opined that there are degrees of complicity. "Other things being equal, it is better for a physician not to be complicit in torture," they assert. "But other things are rarely equal and ... a physician ought sometimes to accept complicity in torture for other moral reasons" (2011, 38). This claim deserves serious consideration.

We might think of complicity in terms of a spectrum. At one end is active participation in torture, where the doctor willingly—under no coercion—assumes a role in the abuse. Somewhere in the middle is passive participation, where the doctor enables torture without initiating or explicitly condoning it. This includes falsifying documents and death certificates to cover up the brutality. At the other end of the spectrum is the refusal to participate. The conscientious doctor will take steps to blow the whistle or otherwise bring the abuse to the attention of authorities. The risks of doing so ought to be weighed into the equation, but, as much as possible, health professionals should take action in reporting this violation of human rights.

The question is to what degree medical personnel should contribute their skills to help victims, even if it means sending them back in, allowing torture to continue. If the victim is seen as a patient, what ought a health caregiver do? Patching up injuries may seem prudent and a caring response, but it can also be seen as playing a contributive role, if not legitimizing, the abuse. This calls for guidelines from professional organizations like the World Medical Association, and the International Council of Nurses.

3.8 The Extent of Doctors' Participation in Torture

The range of doctor involvement is staggering. And they are not just the acts of one country or one detention center—it's a global problem. Doctors who enable and legitimize torture by their participation or by keeping silent seem unclear as to where the moral boundaries should be drawn. Health professionals helped develop, implement and justify torture. They monitored interrogation techniques to determine their effectiveness. This turned detainees into human subjects without their consent and, according to Physicians for Human Rights, bordered on unlawful experimentation (2009, 4). Their presence helps dignify torture, or at least adds credibility.

A central concern is whether short-term gains take precedence over long-term goals. To answer that, we need to determine how best to make that call. Not only do we need criteria to make an assessment, we also need the moral fortitude to admit when mistakes are made and when to turn back from an ill-chosen path.

We need more leaders like Marine Brigadier General Michael Lehnart, who acknowledged that, "I think we lost the moral high ground." As he pointed out, "For those who do not think much of the moral high ground, that is not that significant. But for those who think our standing in the international community is important, we need to stand for [defensible] values. You have to walk the walk, talk the talk" (Perry 2009).

Walking the walk means taking a stronger role in addressing torture and speaking out about the barbaric actions medical personnel are party to. Their role in such abuse as "enhanced interrogation," solitary confinement, and force-feeding warrants stiffer sanctions, given how wide is the circle of participants.

Channels should be in place for educating health professionals regarding signs of torture and ways they could be pressured to participate or keep silent.

3.9 There Are Various Ways to Lose One's Bearings

One way in which doctors enable torture is by deception, such as falsifying records. We saw this in the killing of an innocent Afghani cab driver, Dilawar. The documentary *Taxi to the Dark Side* details his death at the hands of interrogators who beat him so viciously over a 24-h period that his legs were "pulpified" (Shamsi 2006). Steven H. Miles reports that the doctor stated on the death certificate that Dilawar died of natural causes, thereby covering up the crime (Miles 2009; Teays 2008).

In light of the fact that no reasonable person, much less a doctor, could possibly mistake pulpified legs for anything remotely natural, the deception was also self-deception. In addition, it suggests that the doctor falsifying Dilawar's death certificate did not think anyone would notice or raise questions or that violating international medical codes should be prohibitive. As for feelings of guilt and shame, they did not appear to be a factor in the misconduct. These concerns are too important to dismiss.

Steven H. Miles has done extensive work on doctors who abandon their fiduciary duties to patients, thus enabling torture. He criticizes what we have seen with members of the medical profession like Jessen and Mitchell who helped design and implement abusive interrogations. Plus, doctors have monitored interrogation techniques to determine their effectiveness, thus participating in unethical medical experimentation. "Clearly doctors went along with misrepresentations to keep things under cover," observes Psychiatry Professor Leo Eisenberg (Brown 2005; Teays 2008).

Bioethicist Jonathan H. Marks (2005) says the evidence "makes clear that medical personnel were not simply gatekeepers." There is evidence, for example, that they reviewed detainee medical records to find "weak spots", such as a severe phobia of the dark—and that they advised interrogators to exploit detainees' fears and induce extreme stress (Marks 2005).

3.10 There Are Still Lessons to Learn

Loveluck and Zakaria of *The Washington Post* report on the participation of doctors in torture in Syria in the last few years, pointing out that,

> Investigators say that testimony and documentation from Syria's military hospitals offer some of the most concrete evidence to date of crimes against humanity that could one day see senior government figures tried in court. "We were swept into a system that was ready for us. Even the hospitals were slaughterhouses" …
>
> Medicine has been used as a weapon of war since the earliest days of the uprising, when pro-government doctors performed amputations on protesters for minor injuries. Military hospitals across Syria have long set aside wards for prisoners. But since 2011, these have been packed with men left starving and broken by the conditions they have already endured (Loveluck and Zakaria 2017).

Helen McColl, Kamaldeep Bhui, and Edgar Jones rightfully contend that we need to apply sanctions or other measures to emphasize the importance of taking a strong moral stance when it comes to torture. They argue that,

It is vital that national and international medical organizations recognize and prevent medical complicity in torture. Although many of these organizations endorse the ethical codes and human rights instruments that prevent medical complicity, only few of these organizations implement action against medical complicity (2012).

The authors offer two examples of medical organizations facing the decision whether to take steps beyond denouncing and/or prohibiting the active or passive participation in torture. These are the Chilean Medical Association and the South African Medical Association. The first is Chile after the end of the Pinochet regime, during which hundreds of citizens were tortured. The Chilean Medical Association investigated and expelled a number of doctors who were involved in torture—sending a clear message to medical personnel.

The second is South Africa, where two doctors were punished after failing to treat or report the injuries of anti-apartheid activist Steve Biko, who died in police custody from torture-related injuries. "However, the doctors were only punished eight years later, during which time the South African Medical Association failed to respond as it should have done." This led to the withdrawal of the South African Medical Association from the World Medical Association, to pre-empt their expulsion (McColl et al. 2012).

It is crucial for medical organizations to respond to doctors failing to uphold their duties to their patients. This is especially important with regard to such moral failings as torture. As was the case in Chile and South Africa, whether or not to sanctions are put into effect sends a message that reverberates around the world. It is not just the organization's members who are called to act responsibly in caring for their patients, but it reminds all other health professionals why professional codes and ethical principles are important as guidelines for behavior.

Let's see what happens when secrecy comes into play and the challenges it raises for keeping an ethical foundation in place.

3.11 Secrecy

Torture victims are a vulnerable population when it comes to human rights. Lack of due process and restricted access by watchdog agencies like the International Red Cross add to the problem. Isolated from public view, they are at the mercy of their guards and interrogators. Systemic issues are more likely to go unchecked when independent monitors or review boards (IRBs) are not part of the equation.

The war on terror has been shrouded in so much secrecy that much of what has transpired has been withheld from discussion and evaluation. It has also had a detrimental impact since attorneys seek information regarding their clients on a larger scale, one with global repercussions. Countries have colluded with one another to send or receive suspects who are at a loss to know what lies ahead for them and when it will end.

Given this is all takes place under the cover of **"state secrets,"** it is short of impossible to respond to the policies and procedures being used. As a result, the lack of transparency leaves detainees in a hellish limbo. They may have no idea how long it will be until they see the light of day or who, if anyone, will come to their aid. The many testimonies indicate how extensive has been degrading treatment and torture. Furthermore, since most of them have not been charged with the crime it is hard to defend themselves or know the evidence against them.

Confessions are unreliable. They may only be attempts on the part of suspects to end the abuse. Anything to try to stop the torture. We see this with Khalid Sheikh Mohammed's admission that, "I gave a lot of false information to satisfy what I believed the interrogators wished to hear in order to make the ill-treatment stop" (Danner 2007, 23). Torture thus functions more a form of punishment than a way to get actionable intelligence.

3.12 The Questionable Effectiveness of Torture

Psychiatry Professor Edmund Howe points out that "harsher methods" such as torture are simply not the best way to extract information. He asserts that:

> The net effect, in terms of consequences, must be assessed overtime and in its broadest terms. For instance, even if harsher methods do save more lives in the short-term, they may also result in more lives been lost over the longer run. The use of harsher approaches may, for instance, cause greater animosity and, indeed hatred in others for generations to come. In net effect, this may result in a greater loss of lives (2009, 77).

Bioethicist Jonathan H. Marks argues along a similar vein, reinforcing Howe's point and pointing out the significance of international laws. He says,

> The deployment of aggressive interrogation strategies … violate fundamental norms of international human rights law and the laws of war … The experts made clear time again that neither torture nor aggressive interrogation is the best way of producing reliable intelligence. Rather, an approach that builds rapport is much more effective (2009, 21–22).

Not enough credence has been given to long-term consequences of abusive practices. However, such widespread harm casts a shadow that can last in the collective memory for years to come. That's one reason for Sabrina Harmon's observation that what happened at Abu Ghraib was, effectively, a way to recruit more terrorists rather than deter them, as noted in the documentary *Standard Operating Procedure*.

3.13 Rendition

One of the centerpieces of secrecy is a program known as "rendition." Those undergoing it are called "rendered." Rendition involves the transfer of terrorist suspects to detention and interrogation in countries where federal and international legal safeguards do not apply, as the American Civil Liberties Union (ACLU) asserts.

The ACLU (2018) reports: "Foreign nationals suspected of terrorism have been transported to detention and interrogation facilities in Jordan, Iraq, Egypt, Diego

Garcia, Afghanistan, Guantánamo, and elsewhere." Even Lithuania has played a part in rendition, reports *The Washington Post* (Miller et al. 2014).

Mark Danner (2007) lists other countries in the secret system of rendition (2007, 2). They include Thailand, Morocco, Poland, and Romania into which, "at one time or another, more than 100 prisoners disappeared." This is obviously not an incidental number of people who have been subjected to this practice.

In the words of former CIA agent Robert Baer: "If you want a serious interrogation, you send a prisoner to Jordan. If you want them to be tortured, you send them to Syria. If you want someone to disappear—never to see them again—you send them to Egypt" (ACLU 2018).

3.14 The CIA "Black Sites" Offer the Ultimate Secrecy

The Washington Post reports that, at its height, the CIA program included secret prisons in "locations that are referred to only by color-themed codes in the report, such as 'COBALT,' to preserve a veneer of secrecy." Furthermore, the color-coding of locations of rendition and establishment of the "black sites" was part of a broader transformation into a paramilitary force. This was one with new powers to capture prisoners, disrupt plots, and carry out targeted killings (= assassinations) (Miller et al. 2014).

These black sites took secrecy to a whole new level and made accountability all that more difficult to achieve. It also meant the one phone call and access to legal counsel was a pipe dream. Those sent off—rendered—to black sites effectively disappeared off the planet, as far as their friends and family were concerned. They were dead to the outside world until such time that they were released—if they were released. Black sites, black holes.

Human Rights Attorney Stephanie Erin Brewer and Social Psychologist Jean Maria Arrigo remark on how ominous and alarming this is. Specifically, they point out that the amount of secrecy is simply sinister:

> Although protests against conditions in US detention centers occupy the spotlight, an entire network of US counterterrorism operations— arguably more sinister and responsible for more frequent and severe cases of torture— remain largely in darkness … Indeed, the environment of total secrecy in which these operations play out evokes the concern expressed by some … [that] more egregious treatment of detainees is conducted elsewhere [than Guantanamo Bay] (2009, 2).

Another sinister form a black site can take is mobile—thus the use of ships in rendition. Brewer and Arrigo examine this state of affairs, noting that:

> Indeed, the human rights group Reprieve recently revealed that the United States operates "floating prisons " by detaining and interrogating prisoners onboard numerous ships, where physical abuse is reportedly worse than in Guantánamo. The ships are surely staffed with health professionals (2009, 13).

Use of ships to detain and interrogate suspects makes accountability even more of a hurdle than with facilities on the land. Also, there is no pretense of suspects having access to legal counsel. Do the terms "due process" mean anything when it comes to "floating prisons"? How can we tell? We cannot scrutinize what we cannot see. This underscores the importance of human rights.

3.15 Secrecy Undermines Human Rights

Brewer and Arrigo emphasize the essential role that secrecy plays in undermining human rights. This we can see by a liaison officer's perceptions about the advantages of such black sites: "It's so nice to be secret ... So secret that most of the military or government have no idea where [you] are. No rights, human or otherwise have to be dealt with" (Brewer and Arrigo 2009, 13). Secrecy unleashes power; namely heightened power over the suspect.

Along with human rights violations, such secrecy creates conflicts with the Patient's Bill of Rights: "If you have severe pain, an injury, or sudden illness that convinces you that your health is in serious jeopardy, you have the right to receive screening and stabilization emergency services whenever and wherever needed, without prior authorization or financial penalty" (Emergency Services Section, Patient's Bill of Rights).

Let's also note the harm goes beyond the individual level of those sent off to another country or to CIA-run black sites on land or water to face brutal interrogations. It extends to families, communities, workplaces, and neighborhoods. Once people are swept up into the rendition system, loved ones may have no idea what happened to them, where they are, how they can be contacted, and when in the world they will ever reappear.

For example, the German used-car salesman and innocent victim of mistaken identity Khalid el-Masri was sent off to Syria. There he faced almost a year of torture and solitary confinement in a coffin-like space. He had no way to contact his family. As a result, his wife and children didn't know if he was alive or dead—clearly a traumatic experience with long-term complications. Furthermore, depending upon his role in the community, others suffered and relationships were disrupted, with a long road back to normalcy.

3.16 Conclusion

The secrecy enveloping torture and the use of rendition makes it difficult on the part of concerned citizens and professional organizations to to assess the policy and effect change. On a global level we might wonder how the practice got approved and why other countries agreed to play a role. The moral failure in countries agreeing to support this practice leaves a long shadow. What were they thinking?

Fortunately, some countries have ended up having second thoughts about their participation. For example, Italy.

3.16 Conclusion

> In a landmark ruling, an Italian judge on Wednesday convicted a base chief for the Central Intelligence Agency and 22 other Americans, almost all C.I.A. operatives, of kidnapping a Muslim cleric from the streets of Milan in 2003 …
>
> Italian prosecutors had charged the Americans and seven members of the Italian military intelligence agency in the abduction of Osama Moustafa Hassan Nasr, known as Abu Omar, on Feb. 17, 2003. Prosecutors said he was snatched in broad daylight, flown from an American air base in Italy to a base in Germany and then on to Egypt, where he asserts that he was tortured (Donadio 2009).

All the Americans involved were tried in absentia and are considered fugitives. That Italy took steps to place charges against CIA agents for what they did in their country stands out as significant in bringing to light such morally questionable activities.

Britain has also stepped up to the plate. As Richard Perez-Pena of *The New York Times* reports:

> Britain's intelligence services tolerated and abetted "inexcusable" abuse of terrorism suspects by their American counterparts, according to a report released by Parliament on Thursday that offers a wide-ranging official condemnation of British intelligence conduct in the years after the Sept. 11, 2001, attacks ….
>
> The committee documented dozens of cases in which Britain participated in sending suspects to other countries that were known to use torture or aided others in doing so — a practice known as rendition (2018).

Lamentably, "Prime Minister Theresa May accepted the findings but described the intelligence services' moral lapses as a result of bad preparation, rather than of malice" (*The New York Times* 2018). That said, the fact that countries are starting to question their complicity in rendition and torture is to be commended. As more information comes to the surface, there's more reason to hope human rights will prevail over secrecy.

Lieutenant Commander Charles Swift, attorney for Salim Ahmed Hamdan in the US Supreme Court case of *Hamdan v. Rumsfeld* raises a red flag about the way secrecy puts us all at risk. His concerns should prod us to a serious reassessment of the status quo:

1. If you can be the executioner without telling anyone about it … what's left?
2. If I can decide the reasons you will be held in jail for the rest of your life and I alone get to know them and I don't have to tell anyone, what's left?
3. When laid bare, their argument is, there is no limit on presidential power. The president gets to decide his power and no one else.
4. When things are secret, we don't have to be responsible. We can live our lives without taking responsibility for our country. But we are all responsible. We get the country we deserve, because we chose it.

Opening the door on torture is akin to letting the Alien on the ship—once aboard, it grows and multiplies. And we can no longer be sure who or what is in control.

As citizens, doctors and nurses may condone brutal interrogation methods—but, as health caregivers, the emphasis shifts. The subjects are not just "detainees," they are patients. It is crucial that those representatives of our public health system steer a clear moral course. Of paramount importance is their fiduciary duty to patients, even suspected terrorists.

They must not lose sight of the ethical codes and values of the public health system. When facing abusive practices, doctors and other caregivers need to be clearly supported—e.g., by the armed forces, government agencies, medical associations and professional groups—as well as the general public and the countries in which they live and practice. The widespread indifference to the fate of detainees has posed considerable obstacles in addressing the problems they have faced. On a global scale, as we have seen, that calls for examination and action.

Works Cited

American Civil Liberties Union. *Fact Sheet: Extraordinary Rendition*, https://www.aclu.org/other/fact-sheet-extraordinary-rendition. Retrieved 10 June 2018.

Annas, George J., Sondra S. Crosby, and Leonard H. Glantz. 2013, July 11. Guantanamo Bay—A Medical Ethics Free Zone? *New England Journal of Medicine.* https://www.nejm.org/doi/full/10.1056/nejmp1306065. Retrieved 20 March 2019.

Bellamy, Alex J. 2006. No Pain, No Gain? Torture and Ethics in the War on Terror. *International Affairs* 82. https://onlinelibrary.wiley.com/doi/abs/10.1111/j.1468-2346.2006.00518.x. Retrieved 10 June 2018.

Brewer, Stephanie Erin, and Jean Maria Arrigo. 2009. Preliminary Observations on Why Health Professionals Fail to Stop Torture in Overseas Counterterrorism Operations. In *Interrogations, Force-Feedings, and the Role of Health Professionals*, ed. Ryan Goodman and Mindy Jane Roseman. Cambridge, MA: Human Rights Program at Harvard Law School.

Brown, Kevin. 2005, November 7. Medical Ethicist to Discuss Prisoner's Health and Human Rights. *University [of Michigan] Record.* www.ur.umich.edu/0506/Nov07_05/05.shtml. Retrieved 10 June 2018.

Clark, Josh. 2011, May 11. Should Doctors Assist Torture Prisoners? *Stuff You Should Know.* https://www.stuffyoushouldknow.com/blogs/should-doctors-assist-torture-prisoners.htm. Retrieved 10 June 2018.

Danner, Mark. 2007, February. ICRC Report on the Treatment of 14 "High Value Detainees" in CIA Custody by the International Committee of the Red Cross. *U. S. Torture: Voices from the Black Sites Sites* 56 (6), April 9, 2009. http://www.nybooks.com/media/doc/2010/04/22/icrc-report.pdf. Retrieved 10 June 2018.

Dayan, Colin. 2007. *The Story of Cruel and Unusual*. Cambridge, MA: MIT Press.

Fay, Maj. Gen. George. 2004, August 25. Fay Report: Investigation of 205th Military Intelligence Brigade's Activities in Abu Ghraib Detention Facility. *The Torture Data Base.* https://www.thetorturedatabase.org/document/fay-report-investigation-205th-military. Retrieved 10 June 2018.

Donadio, Rachel. 2009, November 4. Italy Convicts 23 Americans for C.I.A. Renditions. *The New York Times.* https://www.nytimes.com/2009/11/05/world/europe/05italy.html. Retrieved 10 June 2018.

Fink, Sheri. 2017, August 17. Settlement Reached in C.I.A. Torture Case. *The New York Times.* https://www.nytimes.com/2017/08/17/us/cia-torture-lawsuit-settlement.html. Retrieved 10 June 2018.

History Commons. 2009, April 18. Doctors, Medical Ethicists Horrified at News of Medical Professionals' Participation in Torture. http://www.historycommons.org/context.jsp?item=a041809torturemedethics#a041809torturemedethics. Retrieved 10 June 2018.

Howe, Edmund. 2009. Further Considerations Regarding Interrogations and Force-feeding. In *Interrogations, Force-Feedings, and the Role of Health Professionals*, ed. Ryan Goodman and Mindy Jane Roseman. Cambridge, MA: Human Rights Program at Harvard Law School.

Human Rights Watch. 2011, March 8. Zimbabwe: No Justice for Rampant Killings, Torture. *Human Rights Watch*. https://www.hrw.org/news/2011/03/08/zimbabwe-no-justice-rampant-killings-torture. Retrieved 10 June 2018.

Kiley, Kevin C., 2005. Memorandum for Record, The Army Surgeon General. *Army Surgeon General Report*. http://hrlibrary.umn.edu/OathBetrayed/Army%20Surgeon%20General%20Report.pdf. Retrieved 10 June 2018.

Larson, Heidi. 2012, May 27. The CIA's Fake Vaccination Drive Has Damaged the Battle Against Polio. *The Guardian (UK)*. http://www.guardian.co.uk/commentisfree/2012/may/27/cia-fake-vaccination-polio. Retrieved 10 June 2018.

Lepora, Chiara and Joseph Millum. 2011, May–June. The Tortured Patient. *The Hastings Center Report*. 38–47. www.thehastingscenter.org/Publications/HCR/Detail.aspx?id=5360. Retrieved 10 June 2018.

London, L., D. Ncayiyana, D. Sanders, A. Kalebi, and J. Kasolo. 2008, October. Editorial. Zimbabwe: A Crossroads for the Health Professions. *South African Medical Journal* 98: 77–78. www.samj.org.za/index.php/samj/article/view/2735/2138+Rayner+M.+Turning+a+blind+eye%3F&cd=5&hl=en&ct=clnk&gl=us&client=safari. Retrieved 10 June 2018.

Loveluck, Louise and Zakaria Zakaria. 2017, April 2. The Hospitals Were Slaughterhouses': A Journey Into Syria's Secret Torture Wards. *The Washington Post*. https://www.washingtonpost.com/world/middle_east/the-hospitals-were-slaughterhouses-a-journey-into-syrias-secret-torture-wards/2017/04/02/90ccaa6e-0d61-11e7-b2bb-417e331877d9_story.html?utm_term=.3fe6828e0864. Retrieved 10 June 2018.

Marks, Jonathan H. 2005, July–August. Doctors of Interrogation. *The Hastings Center Report* 35 (4). http://www.jstor.org/stable/3528822. Retrieved 10 June 2018.

———. 2007, March–April. The Bioethics of War. *The Hastings Center Report* 37: 41–42. https://onlinelibrary.wiley.com/doi/full/10.1353/hcr.2007.0029. Retrieved 10 June 2018.

———. 2009. Looking Back Thinking Ahead: The Complicity of Health Professionals in Detainee Abuse. In *Interrogations, Force-Feedings, and the Role of Health Professionals*, ed. Ryan Goodman and Mindy Jane Roseman. Cambridge, MA: Human Rights Program at Harvard Law School.

Mazzetti, Mark. 2012, January 28. Panetta Credits Pakistani Doctor in Bin Laden Raid. http://www.nytimes.com/2012/01/29/world/asia/panetta-credits-pakistani-doctor-in-bin-laden-raid.html. Retrieved 10 June 2018.

McColl, Helen, Kamaldeep Bhui, and Edgar Jones. 2012, November. The Role of Doctors in Investigation, Prevention and Treatment of Torture. *Journal of the Royal Society of Medicine* 105 (11): 464–471. https://www.ncbi.nlm.nih.gov/pmc/articles/PMC3526851/. Retrieved 10 June 2018.

McNeil Jr., Donald G.. 2012, July 9. C.I.A. Vaccine Ruse May Have Harmed the War on Polio. *The New York Times*. http://www.nytimes.com/2012/07/10/health/cia-vaccine-ruse-in-pakistan-may-have-harmed-polio-fight.html?pagewanted=all. Retrieved 10 June 2018

Miles, Steven H. 2009. *Oath Betrayed: Torture, Medical Complicity, and the War on Terror*. New York: Random House.

Miller, Greg, Adam Goldman and Julie Tate. 2014, December 9. Senate Report on CIA Program Details Brutality, Dishonesty. *The Washington Post*. https://www.washingtonpost.com/world/national-security/senate-report-on-cia-program-details-brutality-dishonesty/2014/12/09/1075c726-7f0e-11e4-9f38-95a187e4c1f7_story.html?noredirect=on. Retrieved 10 June 2018.

Patient's Bill of Rights. *Global Health Source*. https://www.healthsourceglobal.com/docs/Patient%20Bill%20of%20Rights_merged.pdf. Retrieved 10 June 2018.

Perez-Pena, Richard. 2018, June 28. Britain Abetted U.S. Torture of Terrorism Suspects, Parliament Finds. *The New York Times*. https://www.nytimes.com/2018/06/28/world/europe/uk-torture-terrorism.html?rref=collection%2Ftimestopic%2FExtraordinary%20Rendition&action=click&contentCollection=timestopics&region=stream&module=stream_unit&version=latest&contentPlacement=1&pgtype=collection. Retrieved 10 June 2018.

Perry, T. 2009, September 25. We Lost the Moral High Ground. *Los Angeles Times*.

Shamsi, Hina. 2006, February. Detainee Deaths in US Custody in Iraq and Afghanistan. *Command's Responsibility*. http://lawofwar.org/Command's%20Responsibility%20HRW.htm. Retrieved 10 June 2018.

Sonntag, Jesper. 2008. Doctors' Involvement in Torture. *National Center for Biotechnology Information, Torture* 18 (3): 161–175. https://www.ncbi.nlm.nih.gov/m/pubmed/19491477/. Retrieved 10 June 2018.

Sullivan, Andrew. 2006, June 23. Doctors Got into the Torture Business. *Time*. http://content.time.com/time/nation/article/0,8599,1207633,00.html. Retrieved 10 June 2018.

Teays, Wanda. 2008. Torture and Public Health. In *International Public Health Policy and Ethics*, ed. Michael Boylan. New York: Springer.

The Globe and Mail. 2017, June 8. Doctors Tortured Patients at Ontario Mental-Health Centre, Judge Rules. https://www.theglobeandmail.com/news/national/doctors-at-ontario-mental-health-facility-tortured-patients-court-finds/article35246519/ Retrieved 10 June 2018.

The New York Times. 2017, June 20. Psychologists Open a Window on Brutal C. I. A. Interrogations. *The New York Times*. https://www.nytimes.com/interactive/2017/06/20/us/cia-torture.html. Retrieved 10 June 2018.

The Washington Post. 2004, May 7. Rumsfeld Apologizes to Iraqi Prisoners. http://www.washingtonpost.com/wp-dyn/articles/A8563-2004May7_2.html. Retrieved 10 June 2018.

Wisnewski, J. Jeremy, and R.D. Emerick. 2009. *The Ethics of Torture*. London: Continuum.

Chapter 4
Global Torture

> *I was never threatened with death, in fact I was told that they would not allow me to die, but that I would be brought to the "verge of death and back again."*
> —Khalid Sheikh Mohammed.

Overview
The first unit gives an overview of current realities and issues regarding torture and the involvement of doctors, giving us a sense of the challenges facing the medical profession. In this chapter I focus on global torture. To grasp the range of abusive tactics found in countries around the world, I look at six cases. Each is a picture of brutality; together they testify to a moral failure in terms of human rights.

We start with the Mau Mau's fight in Kenya for independence from Britain in the 1950s. We then turn to torture in a Nazi camp In Chile in the 1970s, where human experimentation was practiced as well. Example number three takes us to torture in the Philippines during the Marcos regime and more recently in 2014–15, where torture was practiced as a form of entertainment. Our fourth example is torture in Uzbekistan. There we find the macabre practice of boiling arms or other body parts. Example number five takes us to recent events in Syria, where victims numbered in the thousands. Our last example, number six, is Abu Ghraib prison in Iraq. As with the "Wheel of Torture" game in the Philippines, Americans abused detainees at Abu Ghraib for their own entertainment.

It is noteworthy that guilt, shame, and regret are starting to bubble to the surface, making it clear that the victims aren't the only ones harmed by torture.

4.1 Introduction

Torture takes place on global scale and the participants have included medical personnel who either passively or actively played a role in the abuse of its victims. Fortunately, at least some of the incidents have come to light and victims may have some success in seeking redress. A brief survey shows the ways in which torture is an international problem and not restricted to one group or country.

We will look at six examples to get a sense of the scale and range of global torture. In some cases, victims number in the thousands, as recently shown in Syria. They range over age, ethnicity, gender, political and religious affiliation.

4.2 First Case: The Mau Mau of Kenya (1950s)

The Mau Mau of Kenya fought for independence from Britain in the 1950s. It was a high price to pay. Lawyers for the Mau Mau asserted that the torture and inhumane treatment was systemic, resulting from policies sanctioned at the highest levels of the British government. In other words, the mistreatment of the Mau Mau was not the action of a few "bad apples" violating human rights. The perpetrators were British officials and prison guards.

Over a million Kenyans were detained and affected. The abuse included sexual assault (and rape by glass bottles), castration, beatings, and being burned alive. The torture was widespread, as acknowledged later via a multimillion dollar settlement in 2013. This included an apology on the part of the British government in which they said they "sincerely regret" the abuses that took place (Leigh Day & Co. 2009).

The shocking thing is that these crimes were committed by British colonial officials—perpetrators who were not recognized or punished for decades. One officer was accused of beating up one of the Mau Mau captives and setting him on fire, killing him. Even though they occurred many years ago, the physical and mental scars remain, reports *The Guardian (UK)* (*Press Association* 2013).

Although the uprising of the Mau Mau who sought independence from Britain was crushed by 1956, the Kenyan administration still had to deal with thousands held in camps. They employed a system of assaults and psychological shocks on detainees to force the compliance of the supporters of the revolt. This included being beaten, forced to the ground and having mud forced into their mouths, and stretched out and castrated with pliers (Casciani 2011).

Neither an apology nor financial compensation could make up for being permanently maimed. The torture of the Mau Mau had long-term consequences, given the viciousness of their treatment.

4.3 Doctors Played a Role in the Abuse

Bioethicist Steven H. Miles, who has written extensively on torture, including that of the Mau Mau, condemns doctors who participated or were complicit in their abuse. He recognizes the pressures and coercion medical personnel may face, but calls for a stronger response on the part of governing bodies. He points out that,

Accountability for physicians who abet torture does not resolve many ethical problems. It does not resolve the issue of whether a physician's fear of being tortured for resisting torture is so well grounded as to excuse their silent complicity with orders. Nevertheless, the justified fear of reprisals [does not excuse] inaction by medical boards and courts and physicians in open societies, such as the United States' clinicians who oversaw waterboarding or British physicians who silently treated castrated Mau-Mau (Miles 2014).

4.4 Second Case: Colonia Dignidad Nazi Camp in Chile (1970s)

Chile under the Pinochet regime was infamous for human rights violations. One of the many was a Nazi camp where hundreds of people were tortured. It was also a facility where human experiments and medical procedures reminiscent of those in Nazi Germany were conducted. Similarly, medical personnel were directly involved. In brief:

Chile's feared national intelligence agency, DINA, was responsible for the torture and killing of thousands of political prisoners after the country's 1973 military coup. It established a secret interrogation centre in Colonia Dignidad, a German sect in a remote region of southern Chile, where Paul Schaefer, a former Nazi army nurse and the leader of the sect, taught the Chileans new and brutal methods of torture. The bodies of scores of his victims were later discovered buried on the grounds of the Colony (*al-Jazeera* 2013).

Horrific medical techniques were put to use. For example, "People would say that they did more macabre things [at Colonia Dignidad], they would take out corneas, take out eyes" *(al-Jazeera* 2013). One who delivered a prisoner for what he thought would be interrogation, reports that,

We left him at the entrance of Colonia Dignidad. At first I thought it was just another Chilean military barracks, because I saw guns, they were wearing uniforms of the Chilean army ... But then I realised, no, this isn't the Chilean army. ... [It was]. Germans, but with Chilean army uniforms .... Hours passed ... suddenly the 'Professor', Paul Schaefer, appeared from another room. ... He made a signal to us with his hands. "Fertig," he said. ... Fertig. It means it's over, it's done. And I [understood] that El Loro, the prisoner, was dead. No one asked any questions, least of all me (*al-Jazeera* 2013).

Why did they bring the prisoner to Colonia Dignidad? "Perhaps because the torture methods were more scientific, people would say that they did more macabre things ... those were the rumours. I do know that they experimented with the prisoners, to see how much pain they could withstand" *(al-Jazeera* 2013). He tells of painful experiments—obviously without the consent of the subject/victim.

One moral failing was the use of injections: "They would inject Pentothal, or they would inject things that weren't normal. ... Prisoners who had survived at least one visit, they told me ... that they underwent much harsher interrogations, trying to test the behaviours to see how much they could resist. They used injections, but also methods that were very cruel as well" (*al-Jazeera* 2013).

Evidently the colony was offered to the new regime and, through the 1970s, it became a place for political dissidents to be brought for torture by Pinochet's secret intelligence agency. Some 350 people were tortured and about 100 were murdered and buried there.

And that's not all. A disturbing speculation is the role of ex-Nazis in the colony:

> It was also believed to have played host to "Angel of Death" Josef Mengele, a Nazi doctor known for his lethal human experiments during the Holocaust. Nazi fugitive Walter Rauff, who invented the portable gas chamber, is also said to have found refuge in the colony. Colonia Dignidad apparently thrived under Pinochet's regime, which offered it tax breaks for being a charitable organization and turned a blind eye when escapees reported abuses" (Learning History 2017).

4.5 Third Case: Torture in the Philippines (2014)

Unless we think torture is a thing of the past, as with the Mau Mau or Colonia Dignidad, we can turn our attention to recent maltreatment. Consider police abuse in the Philippines. This case shows us that detention centers are not the only places where torture occurs. In a bizarre turn of events, Filipino police were inspired to have "fun" by turning torture into a game, a source of entertainment. Unfortunately, the pawns in the game were human beings.

According to Amnesty International, the "Wheel of Torture" was used in December 2014. Police reportedly used electric shock, beatings, burning with cigarettes, and waterboarding. They also used humiliation; for example stripping suspects naked with their genitalia tied to a string—which police officers then pulled (*The New York Times* 2014). Police officers would spin the wheel to decide what punishment to inflict on their victims. Some victims said they were beaten with steel bars or baseball bats as part of the game (Corcoran 2014).

One "game" was named after the Filipino boxer Manny Pacquiao. It involved hanging suspects upside down and punching them (presumably like Pacquiao) supposedly to extract information (*Associated Press in Manila* 2014). Amnesty International considers the conviction of a police officer committing those acts of torture to be a historic ruling that "plants the seed of hope that the tide may be turning against impunity for perpetrators of torture" (Amnesty International 2016a).

This ruling is significant in light of torture in the days of the Marcos regime in which martial law was in place from 1972 to 1981. Amnesty International estimates that during martial law 70,000 people in the Philippines were imprisoned, 34,000 of which were tortured and 3240 were killed. It's hard to comprehend the scope of such brutality.

Methods used included beatings, injections of "truth serum", Russian roulette, electric shock, sexual assault, strangulation, cigar and iron burns and the so-called "water cure," where water is forced through victim's mouth and forced out through beating. In addition to the use of pain, victims were subjected to solitary confinement and threat of death, rape or other forms of vicious abuse, such as removing organs presumably to hide torture (Hapal 2016).

4.6 Torturers and Torture Victims

Filipinos were not the only ones committing torture; they were also torture *victims* at an earlier time. Most notoriously is the use of the so-called water cure as far back as a century ago. The "water cure" has a long history, as detailed by Historian Paul Kramer in *The New Yorker* (2008). Photographs taken in May 1901 reveal its use.

In the battle over the island, American and Filipino forces engaged in a war that costs the lives of hundreds of thousands of Filipinos and approximately 4000 American soldiers. Kramer reports that US forces were committing atrocities including the water cure: "Lay them on their backs … [put a] stick in the mouth and pour a pail of water in the mouth and nose … They swell up like toads. I'll tell you it is a terrible torture" (Kramer 2008).

Recent events have involved the use of waterboarding and water dousing, which are similar to the water cure in being a form of simulated drowning. Such extreme techniques still hold sway. That US President Donald Trump condones the use of such methods of torture indicates how far we have not come. That terrorist suspects Abu Zubaydah and Khalid Sheikh Mohammed were subjected to extensive waterboarding —83 times for Zubaydah and 183 times for Mohammed—indicates the willingness to use brutality in the attempt to extract information.

4.7 Fourth Case: Uzbekistan (2014–2015)

Another recent example of the use of torture is found in Uzbekistan. Some of those seeking asylum in such countries as Russia or Norway are turned back to face torture back home, in Uzbekistan. According to Amnesty International, "In Uzbekistan torture is rife. Those who are returned have their teeth kicked in, their ribs broken, and are beaten into confessions. Their words are used to convict them" (Amnesty International 2016b).

An example of the abuse of refugee seekers sent back from Norway is reported in the following:

> They asserted that security forces had held them incommunicado for long periods of time and subjected them to torture by beatings on the soles of their feet, head and body, applying electric shocks and depriving them of food for up to six days. State-appointed lawyers were aware that their clients had been tortured, but did not try to lodge a complaint about torture, as reported by the International Partnership for Human Rights (2015).

Amnesty International reports that torture and other ill-treatment have become defining features of the Uzbekistan criminal justice system. Evidently this is central to the way Uzbekistani authorities deal with dissent and actual or perceived security threats, as well as the repression of political opponents (International Partnership for Human Rights 2015).

According to Craig Murray, former British ambassador to Uzbekistan, "A handful have emerged from what has been labeled a secret gulag, and have given deeply disturbing accounts of horrific mistreatment" (Teays 2008, 68). Among the examples that Murray noted includes the boiling of an arm, which he cited as a common method of torture employed in Uzbekistan.

4.8 The Lack of Transparency Adds to the Problem

Fred Hiatt of *The Washington Post* reported on concerns about Uzbekistan raised by Human Rights Watch. In particular, Human Rights Watch indicated that "Uzbekistan has become virtually closed to independent scrutiny." Foreign correspondents and human rights monitors generally are not granted visas. No U.N. human rights expert has been allowed in since 2002 (Hiatt 2014). As with other cases where there is a lack of clarity and accessibility in terms of human rights violations, it is important that the international community exert pressure for that to change.

4.9 Fifth Case: Syria (2003–2017)

Syria offers our fifth example of global torture, also a recent one. It is reported that "unimaginable torture" has taken place in Syria's prisons. Stephen Grey reports that,

> In 2003, the State Department, quoting human rights organizations and former prisoners, would describe Syria's torture methods as including: administering electrical shocks; pulling out fingernails; forcing objects into the rectum; beating, sometimes while the victim is suspended from the ceiling; hyper-extending the spine; bending the detainees into the frame of a wheel and whipping exposed body parts; and using a chair that bends backwards to asphyxiate the victim or fracture the victim's spine (2006).

In 2014, *CNN* publicized what they called "gruesome" photos to prove that torture was taking place in Syria under the Assad regime. In a January 2014 *CNN* newscast, Journalist Christiane Amanpour reported on direct evidence of systematic torture and killing. Furthermore, *CNN* told of thousands of photographs indicating the range of abuses: "The bodies in the photos showed signs of starvation, brutal beatings, strangulation, and other forms of torture and killing, according to the report" (Krever and Elwazer 2014).

Human Rights Watch found evidence of widespread torture, starvation, beatings, and disease in Syrian government detention facilities, in addition to mass killings that were documented by 28,000 photographs smuggled out of Syria and brought to public attention in January 2014. *Washington Post* journalists Louise Loveluck and Zakaria Zakaria (2017) point out that:

> In interviews across Lebanon, Turkey and Europe, more than a dozen survivors and army defectors described horrors in Syrian military hospitals across the country for which war crimes lawyers say they have struggled to find a modern parallel .... Everyone gets the "welcome" party, [Mohsen al-] Masri said — a savage beating involving guards and medical staffs wearing white coats over military uniforms. In Hospital 601, the weakest man was pushed to the floor and brutalized first. In the nearby Tishreen Military Hospital ... Mohammed al-Hammoud, said he had seen prisoners dragged down steps by the hair.

One site of torture is now infamous. That is the hospital, known as 601, where it was seen in a collection of photographs showing thousands of skeletal corpses.

> Inside the facility, about a half-mile from Syrian President Bashar al-Assad's palace, sick prisoners are tortured as they lie shackled to beds crammed with dying men, according to Masri and former detainees and military personnel who worked there. Corpses have been piled in bathrooms, outhouses and anywhere else they will fit, then meticulously documented and trucked away for mass burial (Loveluck and Zakaria 2017).

We should also note that Syria was one of the countries used by the CIA for the rendition of terrorist suspects. Not a pretty picture.

4.10 Sixth Case: Abu Ghraib

Thanks to the actions of whistleblower Joseph Darby, brutality at Abu Ghraib prison in Iraq was brought to the attention of his superiors and, ultimately, the world. It was a bombshell, with the mistreatment leaving dire consequences in terms of the moral standing of the United States. Not unlike the Philippine police having fun with the "Wheel of Torture," American soldiers appeared to take pleasure in the abuse and degradation of detainees.

The abuse was extensive. In May 2004 Major General Antonio M. Taguba issued a report that detailed the range of detainee mistreatment. It prompted a Senate Armed Services Committee hearing, having documented a systemic problem: military personnel had perpetrated "numerous incidents of sadistic, blatant, and wanton criminal abuses," as Maj. Gen. Taguba asserts" (2004, 16). In looking back in 2014 on the effect of the report, he reflects:

> The findings, along with what became infamous images of abuse, caused a stir and led to prosecutions. The inquiry shed light on our country's trip to the dark side, in which the United States government engaged in an assault on American ideals, broke the law and in so doing strengthened our enemies ...
>
> It was clear to me in 2004 that the United States military could not be the institution it needed to be as long as it engaged in and tolerated abuse. But the military's path to accountability was a long one, and its leaders hardly welcomed oversight (Taguba 2014).

He asserts that the intentional abuse of detainees by military police personnel included the following acts:

> Punching, slapping, and kicking detainees; jumping on their naked feet; Videotaping and photographing naked male and female detainees; Forcibly arranging detainees in various sexually explicit positions for photographing; Forced nudity for days at a time; Forcing naked male detainees to wear women's underwear; Forcing groups of male detainees to masturbate while being photographed and videotaped; Arranging naked male detainees in a pile and then jumping on them; Positioning a naked detainee on a MRE [Meals Ready to Eat] Box, with a sandbag on his head, and attaching wires to his fingers, toes, and penis to simulate electric torture; Placing a dog chain or strap around a naked detainee's neck and having a female Soldier pose for a picture; A male MP guard having sex with a female detainee; Using military working dogs (without muzzles) to intimidate and frighten detainees, and in at least one case biting and severely injuring a detainee; Taking photographs of dead Iraqi detainees (Taguba 2004).

You can see from the long list that vicious actions at Abu Ghraib were a combination of physical as well as mental and psychological torment. The latter effectively humiliated the detainees and made their lives miserable on a daily basis—and reinforced their helplessness. *The Guardian (UK)* reports further helplessness for female detainees subjected to "systemic abuse and torture," such as being raped by US guards and forced to strip naked in front of men (Harding 2004).

Maj. Gen. Taguba suggested that his being monitored and eventually asked to resign reflected a lack of receptivity on the part of the military to his report. His documentation—an exposé of cruelty—revealed serious and systemic problems.

As for the motivation for their behavior, "Sergeant A" of the 82nd Airborne revealed: "Some days we would just get bored so we would … make them get in a pyramid ['them' being naked detainees]." He adds, "We did that for amusement" (Teays 2008). The barbaric treatment of detainees in our names should trouble us all.

> In April 2004, photos showing abuse of prisoners at the prison emerged. The images showed US personnel intimidating and threatening prisoners with dogs, with Iraqis hooded, naked, and forced into strange formations. Allegations in later lawsuits filed by Iraqis also alleged physical and sexual abuse, electric shocks, and the conducting of mock executions … "It's hard for me to say this," [Taleb al-Maleji] exclaims, biting his lip as he describes another incident. "We were totally naked and they were beating us with sticks on our genitals" (Arraf 2013).

Just look at the photos that were made public—the viciousness that was unleashed, and the gleefulness in treating the "enemy" with such brutality. There are a number of websites posting photos of torture; see for example *The Washington Post, Salon,* and *Harpers.*

Some trivialize it, as did Guy Womack, attorney for M.P. reservist Charles Graner, who was convicted of a number of charges, including assault and battery at Abu Ghraib. In his opening statement at Graner's court martial Womack said of the piling up of naked detainees, "Don't cheerleaders all over America make pyramids every day?" He added: "It's not torture" (Sturcke 2005). That claim merits assessment, given how degrading was the treatment.

The Department of Defense CID report reveals others who found abuse and torture a source of entertainment. It's hard not to label this "sadistic." For example, in section 1A of Abu Ghraib there are tack marks on the wooden wall to indicate how many stitches detainees received after being abused, particularly by dog bites, notes Guy B. Adams (2006) citing a CID interview by the Department of Defense.

The perpetrators were not just soldiers seeking to "toughen up" detainees in preparation for interrogation. Robert Jay Lifton points out that doctors had a role to play:

> A May 22 [2014] article on Abu Ghraib in *The New York Times* states that "much of the evidence of abuse at the prison came from medical documents" and that records and statements "showed doctors and medics reporting to the area of the prison where the abuse occurred several times to stitch wounds, tend to collapsed prisoners or see patients with bruised or reddened genitals."

> According to [*The New York Times*], two doctors who gave a painkiller to a prisoner for a dislocated shoulder and sent him to an outside hospital recognized that the injury was caused by his arms being handcuffed and held over his head for "a long period," but they did not report any suspicions of abuse … A nurse, when called to attend to a prisoner who

was having a panic attack, saw naked Iraqis in a human pyramid with sandbags over their heads but did not report it until an investigation was held several months later (Lifton 2004).

The circle of misconduct is wider than has been given its due. Too many people lost their moral bearings at Abu Ghraib. And few did the right thing, the humane thing.

And let us not forget the ghoulish grins on abusive guards at Abu Ghraib in the photos first released in 2003. Susan Sontag (2004) observes, "Looking at these photographs, you ask yourself, how can someone grin at the sufferings and humiliation of another human being?" Good question. Over time some of those grinning have had second thoughts—and remorse. We see this with some of the ex-military who served at Abu Ghraib, Tony Lagouranis and Jeremy Sivits.

Lagouranis looks back on what he saw and what he did. He admitted, "I lost my mind a little bit. Panic attacks, anxiety, insomnia, nightmares. I was shaking all the time. Plus I was really angry." In sum, "I was a mess" (Conroy 2007). He realized that he fell short on the moral scale and all the confessions in the world won't take that away. Lagouranis reflects:

> It takes a unique clarity to stand up and say what everyone thinks is so normal is actually abhorrent. I think I did well under the circumstances, but no one reported what they should have when they should have—including me.
>
> I saw barbaric traits begin to seep out of me and other good and respectable people—good Americans who never should have been put in that position to begin with. They have two choices—disobey direct orders or become monsters. It's a lonely road when everyone else is taking the other one (Conroy 2007).

"What happened at the prison was a horrible thing," reflects Jeremy Sivits, who served a year in prison for his role in the mistreatment of detainees at Abu Ghraib. "I'm like: 'Yeah. That was me. That isn't who I am today. I'm a different person'" (McKelvy 2018). We can only hope.

Furthermore, we can only hope that accountability goes beyond the individual level. As Adams, Balfour, and Reed ask:

> If administrative evil means that people inflict pain, suffering, and death on others but do so neither knowingly nor deliberately, can they be held responsible for their actions?
>
> We believe the answer is yes, but when ordinary people inflict pain, suffering, and death on others in the course of performing their "normal" organizational or policy role, they often justify their actions by saying that they were just following orders and doing their job. This fact reflects the difficulty of identifying administrative evil and the possibility of missing it altogether or perhaps worse, calling mistakes or misjudgments evil (Adams et al. 2006).

4.11 Conclusion

What we see from our six examples is that torture is an oppressive instrument with little regard for human rights. In many cases a combination of mental, physical, and psychological methods is used to magnify the effects. Such clustering can produce significantly greater damage—and harm—beyond what could be achieved by one procedure. Thus we may see a victim subjected to physical pain as well as mental or psychological techniques like solitary confinement, sleep deprivation, and exploitation of fears and phobias, whereby there are no scars to reveal the abuse that took place.

In all of these cases there is a blatant disregard for the suspect's dignity. Torture reinforces a power differential leaving detainees and prisoners with little, if any, moral status. The guards, interrogators, perpetrators all exercise power over the detainees/prisoners. Power brings control and control brings power. Displays of power often involve degradation and humiliation, as well indignities like forced nudity, force-feeding, and the right to toilet facilities.

With the use of physical pain such as burning or maiming the victim, punishment seems as much of a factor as extracting information. The idea that pulling out fingernails, breaking bones, or cutting off a limb or two will result in "actionable intelligence" seems spurious at best. The record of testimony has made it clear that people will say almost anything—even a false confession—in hopes of bringing the torture to an end.

One dimension of power that can surface is the element of racism or other forms of prejudice. These can escalate an already volatile situation. If someone is an "enemy," we are likely to accord them fewer rights and think they deserve whatever fate they're subject to. This is seen in places like Uzbekistan and Syria, where boiling arms or cutting off limbs is viewed as an option, as are the grotesque experiments at Colonia Dignidad in Chile.

There is a lack of concern for the long-term consequences of brutality; for example with the Mau Mau getting their genitals cut off or the torture games in the Philippines. Consider also the rape and sodomy all too common in prisons and detention centers around the world. Even then there is the recognition of how shameful is the extent of sexual abuse. This is shown in the US Congress allowing the public to see photographs of nude male detainees at Abu Ghraib enduring degrading treatment, but not the abuse of women and children. The public response (revulsion) might be detrimental to the war on terror.

4.12 And Then There Is the Medical Profession

That doctors and other medical personnel are involved even in minor roles or as bystanders raises questions about the medical profession and the need for ethical principles to guide its members. We just have to scratch our heads and wonder how health professionals could lend their support to something so abhorrent. Is there no shame? No feelings of guilt? Or do they just think that the "enemy" does not deserve the respect they accord their patients? Are these "enemies"—most of whom have not been charged, much less convicted, of a crime— not deserving of patient rights and human rights?

We will go into professional perspectives in Chapter Nine and see that the professions set forth guidelines regarding—prohibiting—torture. That those guidelines can fall by the wayside merits attention. Professional organizations need to spell out procedures for dealing with medical personnel who abdicate their fiduciary duties, particularly in such egregious ways as enabling torture. This involves looking at the sorts of sanctions that may be in order.

We need also to consider the appropriate response to doctors who prioritize patriotism over patient care. This means resolving what counts as a legitimate excuse for complicity in torture. Jonathan H. Marks (2005) observes that, if we find doctors' participation in torture morally permissible, we will need to rethink medical training. We would then need "to embrace the interrogation ethos and its institutional sequelae, not just the practice." This would include teaching "interrogation medicine" and "interrogation psychiatry" courses (2005). Is this the path we want to pursue?

In an interview with Rory Kennedy, director of the documentary *Ghosts of Abu Ghraib*, Army reservist Israel Rivera saw harms to all participants, not just the victims. "It was a door that I was afraid to walk through," he said. "If you walk through it, at which point do you say it's enough? What's cruel enough? How do you get back from that? And I was afraid that I wouldn't come back, that I would get lost" (Kennedy 2007).

All too many have gotten lost, have crossed a moral boundary, leaving them tainted with the vices they were party to. As for, "How do you get back from that?" the truth, unfortunately is, that some never do. And so we have our work cut out. This calls for work on all three levels—that of individual doctors who must keep their ethical duties before them, that of professional organizations providing guidelines and oversight, and that of countries working together on a global scale to ensure human rights are recognized and sustained.

Works Cited

Adams, Guy B. and Danny L. Balfour and George E. Reed. 2006, September–October. Abu Ghraib, Administrative Evil, and Moral Inversion: The Value of "Putting Cruelty First". *Public Administration Review* 66 (5): 680–693. http://www.jstor.org/stable/3843897. Retrieved 10 June 2018.

al-Jazeera. 2013, December 15. Tales of Torture: A Former Member of Chile's National Intelligence Agency Describes Some of the Methods Used Against Political Prisoners. https://www.aljazeera.com/programmes/aljazeeracorrespondent/2013/10/tales-torture-2013103081121394171.html. Retrieved 10 June 2018.

Amnesty International. 2016a, April. Philippines: Historic Ruling on Police Torture Following Amnesty International Campaign. https://www.amnesty.org/en/latest/news/2016/04/philippines-historic-ruling-on-police-torture-following-amnesty-international-campaign/. Retrieved 5 June 2018.

———. 2016b, April. Uzbekistan: Fast Track to Torture. https://www.amnesty.org/en/latest/campaigns/2016/04/uzbekistan-stop-torture/. Retrieved 5 June 2018.

Arraf, Jane. 2013, July 15. The Scars of Abu Ghraib. *al-Jazeera*. https://www.aljazeera.com/indepth/features/2013/07/2013715113237109200.html. Retrieved 5 June 2018.

Associated Press in Manila.2014, January 28. Philippines Police Played 'Wheel of Torture' Game to Extract Information. *The Guardian (UK)*. https://www.theguardian.com/world/2014/jan/28/philippines-police-wheel-of-torture-game. Retrieved 10 May 2018.

Casciani, Dominic. 2011, April 12. British Mau Mau Abuse Papers Revealed. *BBC*. https://www.bbc.com/news/uk-13044974. Retrieved 5 June 2018.

Conroy, John. 2007, March 2. Tony Lagouranis—Confessions of a Torturer. *Chicago Reader.* www.chicagoreader.com/features/stories/torture. Retrieved 5 June 2018.

Corcoran, Kieran. 2014, January 28. Philippine Police Officers 'Used Wheel of Torture To Abuse Suspects' With One Punishment Named After Boxer Manny Pacquiao. *Daily Mail (UK)*. http://www.dailymail.co.uk/news/article-2547702/Philippine-police-officers-used-wheel-torture-abuse-suspects-one-punishment-named-boxer-Manny-Pacquiao.html. Retrieved 5 June 2018.

Grey, Stephen. 2006. *Ghost Plane: The True Story of the CIA Torture Program*. New York: St. Martin's Griffin.

Hapal, Don Kevin. 2016, February 23. Worse Than Death: Torture Methods During Martial Law. *Rappler.* www.rappler.com. Retrieved 4 June 2018.

Harding, Luke. 2004, May 19. The Other Prisoners. *The Guardian (UK)*. https://www.theguardian.com/world/2004/may/20/iraq.gender. Retrieved 10 June 2018.

Hiatt, Fred. 2014, November 30. The Tyranny You Haven't Heard of. *The Washington Post.* https://www.washingtonpost.com/opinions/fred-hiatt-an-oppressive-central-asian-regime-hides-in-plain-sight/2014/11/30/eab84368-74d8-11e4-9d9b-86d397daad27_story.html?utm_term=.89ec885daf9e. Retrieved 10 June 2018.

International Partnership for Human Rights. 2015, September 23. Endemic Torture in Uzbekistan. http://iphronline.org/endemic-torture-in-uzbekistan20150923.html. Retrieved 10 June 2018.

Kennedy, R. 2007. (Producer and director). *Ghosts of Abu Ghraib* [motion picture].

Kramer, Paul. 2008, February 25. The Water Cure. *The New Yorker.* https://www.newyorker.com/magazine/2008/02/25/the-water-cure. Retrieved 10 May 2018.

Krever, Mick and Schams Elwazer. 2014, January 20. Gruesome Syria Photos May Prove Torture by Assad Regime. *CNN.* http://www.cnn.com/2014/01/20/world/syria-torture-photos-amanpour/index.html. Retrieved 5 June 2018.

Learning History. 2017, May 14. Colonia Dignidad: Nazi Torture Camp. https://www.learning-history.com/colonia-dignidad. Retrieved 10 June 2018.

Leigh Day & Co. 2009, October 22. Veterans Present Evidence of Torture to British Government. *Leigh Day (UK)*. https://www.leighday.co.uk/News/2009/October-2009/Mau-Mau-veterans-present-evidence-of-torture-to-Br. Retrieved 5 June 2018.

Lifton, Robert Jay. 2004, October 7. Doctors and Torture. New England Journal of Medicine 351: 1571–1574. http://www.nejm.org/doi/pdf/10.1056/NEJMp048065. Retrieved 10 May 2018.

Loveluck, Louise and Zakaria Zakaria. 2017, April 2. The Hospitals Were Slaughterhouses: A Journey into Syria's Secret Torture Wards. *The Washington Post.* https://www.washingtonpost.com/world/middle_east/the-hospitals-were-slaughterhouses-a-journey-into-syrias-secret-torture-wards/2017/04/02/90ccaa6e-0d61-11e7-b2bb-417e331877d9_story.html?utm_term=.3fe6828e0864. Retrieved 10 June 2018.

Marks, Jonathan H. 2005, July–August. Doctors of Interrogation. *The Hastings Center Report* 35 (4): 17–22. http://www.jstor.org/stable/3528822. Retrieved 10 June 2018.

McKelvy, Tara. 2018. I Hated Myself for Abu Ghraib Abuse. *BBC.* https://www.bbc.com/news/44031774. Retrieved 5 December 2018.

Miles, Steven H. 2014, January 22. The New Accountability for Doctors Who Torture. *Health and Human Rights Journal.* https://www.hhrjournal.org/2014/01/the-new-accountability-for-doctors-who-torture/. Retrieved 5 June 2018.

Press Association. 2013, June 6. UK to Compensate Kenyans Mau Mau Torture Victims. *The Guardian (UK)*. https://www.theguardian.com/world/2013/jun/06/uk-compensate-kenya-mau-mau-torture. Retrieved 5 June 2018.

Sontag, Susan. 2004, May 23. Regarding the Torture of Others. *New York Times.*

Sturcke, James 2005, January 11. Abu Ghraib Inmates 'Like Cheerleaders'. *The Guardian (UK)*. https://www.theguardian.com/world/2005/jan/11/iraq.usa. Retrieved 10 June 2018.

Works Cited

Taguba, Maj. Gen. Antonio M. 2004, May. Taguba Report: AR 15-6 Investigation of the 800th Military Police Brigade. https://fas.org/irp/agency/dod/taguba.pdf. Retrieved 10 June 2018.
———. 2014, August 5. Stop the C.I.A. Spin on the Senate Torture Report. *The New York Times.*
Teays, Wanda. 2008. Torture and Public Health. In *International Public Health Policy and Ethics*, ed. Michael Boylan. Dordrecht: Springer.
The New York Times. 2014, December 3. Amnesty Report Condemns Police Torture in Philippines. https://www.nytimes.com/2014/12/04/world/asia/report-condemns-police-torture-in-philippines.html. Retrieved 5 June 2018.

Part II
Boundaries of Torture

Chapter 5
Boundaries of Torture

> *The therapeutic mission is the profession's primary role and the core of physicians' professional identity. If this mission and identity are to be preserved, there are some things doctors must not do.*
> —M. Gregg Bloche and Jonathan H. Marks.

Overview

In Part II we look at the boundaries of torture and the ethical issues they bring to light. In this chapter I employ Nel Noddings' notion of the three components of evil as a framework of analysis. Each one is applied to torture, recognizing that its aspects of physical pain and mental and psychological suffering are set out by the UN Convention Against Torture in 1984. One change over the years is the increasing use of abusive practices which leave no scars or other physical evidence. Examples are sleep deprivation, isolation, threats to family members, and forced standing. For perpetrators, such tactics offer significant advantages.

We will consider how things can escalate and lead to further abuse. This is referred to as "force drift." In conjunction, we will look at two dimensions of pain—physical and mental pain. An instance of the latter is a "mind virus," where the victim is made to feel responsible for his or her maltreatment. To complicate the matter, clustering techniques may be put to use, thereby ratcheting up the suffering on the part of the victim.

Another concern we look at here is the role language has played in allowing inhumane tactics to be authorized. For example, replacing the term "prisoner" with "detainee" stripped away a host of protections and left "detainees" vulnerable to all sorts of mistreatment.

5.1 Introduction

Nel Noddings characterizes evil as having three components— pain, separation, and helplessness (1989, 95 and 118). This provides us with a useful framework to approach torture. We tend to think of torture in terms of pain, with the infliction of severe pain as an indicator of torturous acts. That being the case, let's first consider forms of torture having a physical component and then turn to other forms, such as mental and/or psychological.

Torture can have both physical and mental aspects in terms of the experience of pain. That both physical and mental suffering are aspects of torture is acknowledged by the UN Convention against Torture and Other Forms of Cruel, Inhuman or Degrading Treatment or Punishment, which was instituted in 1984. We see this in the following definition:

> For the purposes of this Convention, the term "torture" means any act by which severe pain or suffering, whether physical or mental, is intentionally inflicted on a person for such purposes as obtaining from him or a third person information or a confession, punishing him for an act he or a third person has committed or is suspected of having committed, or intimidating or coercing him or a third person, or for any reason based on discrimination of any kind, when such pain or suffering is inflicted by or at the instigation of or with the consent or acquiescence of a public official or other person acting in an official capacity (Adams et al. 2006, 684).

Note also Article 3 of UN Resolution 37:

> It is a gross contravention of medical ethics, as well as an offense under applicable international instruments, for health personnel, particularly physicians, to engage, actively or passively, in acts which constitute participation in, complicity in, incitement to or attempts to commit torture or other cruel, inhuman or degrading treatment or punishment.

5.2 Where Do We Find Torture—And By Whom?

There are many settings in which torture takes place. Most of the examples in this book focus on the degradation and abuse of detainees. However, brutality is across a wider spectrum and doctors are in the thick of it. The participation of medical personnel is not merely the result of dual loyalties in which they are torn between their professional duties and those related to combating terrorism. For some, acts born of patriotism take precedence—and the consequences leave a lot to be desired.

For example, Journalist Sophie Arie (2011) reports that, "In Italy, doctors and nurses at the prison where G8 protesters were detained in 2001 have been accused of suturing some detainees without anesthesia, leaving them undressed for long periods of time, and hitting and beating them." It's hard to fathom medical professionals exhibiting such callousness and tossing their fiduciary duties to the wind. Such actions reveal a moral failure.

5.3 The Infliction of Pain

With this greater picture in mind, let's turn to the infliction of pain in acts of torture. The range of cases illustrates the sorts of pain that torture can cause. On one end of the spectrum is the physical aspect of pain—physical torture— that can spin out of control and escalate. At the other end is the mental aspect of pain—psychological torture—that can cause long-lasting effects, such as PTSD.

First, keep in mind the scope of the problem and the extent to which health professionals are involved. It's not just one country or an assortment of yahoos tossing their ethics to the wind. According to Dan Agin, the British had doctors present at interrogations of IRA members in the 1970s. The French used physicians to assist in the interrogations during the war with Algeria. And so on. Agin points out that, "The list is long, and now history will add Americans of the early 21st century as a people who tortured prisoners. History has put blood on our hands" (2009).

We shouldn't look for monsters in the collection of perpetrators and enablers. The list of physical and mental torture is certainly long, as is the list of those who put them to use. Torture and degrading treatment are not just committed by those with a tenuous hold on morality. Perfectly "normal" citizens and groups of individuals consider torture a viable option.

One reason for the assumption that torture is worthwhile is the view that, if you inflict enough pain, torture works. Evidence does not support this belief. Nevertheless, the popular view is that even the most stalwart individual can be broken and, therefore, torture should be a viable option. That reality means we need to cast a wide net in assessing culpability.

We see, for instance, the excesses of "civilized" countries like England. Speaking of British torture techniques, David Ignatius (2005) asserts that:

> The British put hoods on their IRA prisoners, just as U.S. interrogators have done in Iraq. The British approved other, harsher methods: depriving IRA prisoners of sleep, making them lean against a wall for long periods, using "white noise" that would confuse them.
>
> The clincher for British interrogators was mock execution. The preferred method in the mid-1970s was to take hooded IRA prisoners up in helicopters over the lakes near Belfast and threaten to throw them out if they didn't talk. Sometimes, they actually were thrown out. The prisoners didn't know that the helicopter was only a few yards above the water.

5.4 The "Five Techniques"

In 1972, following an investigation into the treatment of prisoners in Northern Ireland, then-Prime Minister Edward Heath banned the use of hooding, white noise, sleep deprivation, food deprivation and painful stress positions—known as the "five techniques" (British Broadcasting Corporation 2011). His action made it clear that a line had been crossed and it was time to put an end to these practices.

The "five techniques" found a modified version in acceptable tactics on the part of the U.S. For example, the Department of Defense considered the following interrogation techniques to be humane and legally permissible: "isolation for more than 5 months, sleep deprivation lasting 48 to 54 days during which interrogation took place 18 to 20 hours per day, degradation, sexual humiliation, military dogs to instill fear, and exposure to extremes of heat and cold and loud noise for long periods and combinations of these techniques" (Justo 2006, 1462). Unfortunately, the criteria for judging this "humane" by the Department of Defense was not made public.

5.5 The Domain of Pain: Torture on the Physical Plane

The range of torture involving physical pain includes "walling," where the victim's head is slammed into a wall, beatings, sexual assault, being hung by the wrists from the ceiling, short shackling, hypothermia, and bodily injury. In addition to "physical roughing up; sensory, food, and sleep deprivation", detainees were also subjected to a water pit in which they had to stand on tiptoe to keep from drowning (Savage 2004).

There has also been the use of waterboarding (simulated drowning) on high-profile suspects. In many cases (e.g., at Guantanamo Bay), hunger-striking detainees were force-fed and/or subjected to "rectal hydration," both of which cause physical pain and mental distress—and typically involve the participation of health professionals.

Human Rights Watch (2000) cites a recent case of torture from Chechnya in 2000:

> From the time they entered the Chernokozovo facility, when Russian guards would force them to run a gauntlet of guards who would beat them mercilessly, through their stay in cramped and sordid conditions, to the time they were released, detainees had no relief from torment. Fearing identification and possible future retribution, Russian soldiers in Chechnya frequently wore camouflage uniforms with no division patches or pins that would identify them.
>
> Human Rights Watch calls for a full investigation by the Russian authorities of what happened at Chernokozovo in January and February 2000, for those responsible for human rights violations committed there to be brought to justice, and for compensation to be granted to victims or their relatives.

Moreover, the fact the guards abusing the detainees in Chechnya sought to obscure—hide—their identity from their victims makes it clear that they knew what they were doing and that it was wrong.

5.6 The Use of Mental and Psychological Abuse

Physical torment is a central concern. So, too, is mental and psychological suffering. Examples of torture involving mental pain are environmental manipulation such as 24 hour lighting or darkness, piercingly loud music, solitary confinement or being confined to a box, and threats to family members. In addition, the exploitation of phobias (e.g., fear of vermin) contributes to the victim's level of distress. That reality did not prevent doctors from enabling such a practice.

Medical Historian Giovanni Maio (2001) observes that methods of torture have changed, by expanding the types of mental abuse that are used:

> Traditional methods mainly used physical pain, whereas modern torture also involves psychiatric-pharmacological and psychological techniques. Brainwashing is the oldest form of psychological "torture;" other methods include further deprivation (e.g., sleep deprivation), apparent execution, isolation, dark cells, personal threats, and forced observation of others being tortured.
>
> These techniques are used often because they leave no visible evidence of torture; torture today must be impossible to prove, which would not be possible without medical skills.

The Senate Committee Report on Torture (2014) discusses the use of sleep deprivation which, as Maio states, is now "used often." This is because it does not leave marks and certainly would be experienced as disorienting and extraordinarily stressful. As noted by the Senate Intelligence Committee on Torture (2014):

> Sleep deprivation involved keeping detainees awake for up to 180 hours, usually standing or in stress positions, at times with their hands shackled above their heads. At least five detainees experienced disturbing hallucinations during prolonged sleep deprivation and, in at least two of those cases, the CIA nonetheless continued the sleep deprivation … CIA medical personnel treated at least one detainee for swelling in order to allow the continued use of standing sleep deprivation.

In addition, the Inspector General's report describes several forms of abuse not previously reported that, from a medical and legal perspective, constitute torture (Physicians for Human Rights 2009). These include: (1) mock executions and threatening detainees by brandishing handguns and power drills; (2) threatening harm to family members, including sexual assault of females and murder of the detainee's children; and (3) physical abuse, including applying pressure to the side of a detainee's neck resulting in near loss of consciousness (Physicians for Human Rights 2009).

5.7 Justifying Torture with Hypotheticals

We may wonder why such harsh practices are seen as justifiable. The commonly cited argument for employing torture is nuclear terrorism; namely that there is the equivalent of a "ticking bomb" set to go off in some large metropolitan area such as Paris or New York City and time is not on our side.

Members of the pro-torture faction, including such luminaries as Supreme Court Justice Antonin Scalia and US President Trump, think this justifies lifting the restraints against torture. That the hypothetical scenario is merely speculative is not given much weight. Journalist/Editor Matt Ford (2014) explained in *The Atlantic* that,

> The Senate torture report shows how detached this hypothetical scenario is from reality. In the real world, CIA personnel tortured hundreds of detainees, including ones who commit-

ted no crimes. CIA officers and contractors waterboarded detainees, in some cases hundreds of times. CIA medical personnel flooded their orifices with nutrients via plastic tubes for "behavior control." CIA officials denied detainees access to sanitary facilities and forced them to use diapers for humiliation.

They forced detainees to stand on broken ankles. They subjected one to sleep deprivation for 56 hours until he could barely speak and was "visibly shaken by his hallucinations depicting dogs mauling and killing his sons and family." They threatened to murder detainees' children and sexually assault their mothers. They used the taped cries of an "intellectually challenged" detainee to coerce family members. They even shackled one detainee named Gul Rahman, naked, to a concrete floor in a "stress position," where he died of hypothermia (Ford 2014).

No time-bomb ticked as this happened, as Ford points out.

5.8 The Slippery Slope of "Force Drift"

Consider the physical aspects of pain. When looking at torture, Wisnewski and Emerick (2009) observe that the use of force can easily escalate, making it difficult to control. This is the point where "force drift" comes into play and an interrogator can "slide" into coercion. Attributing the term to Psychologist Michael Gelles, they explain how it evolves:

> [The] tendency to slide into coercion was documented by Dr. Michael Gelles, the chief psychologist at the Naval criminal investigation service. Dr. Gelles warned his superiors of what he called a 'force drift': a natural inclination "to uncontrolled abuse when interrogators encounter resistance."
>
> He further warned that, "once the initial barrier against the use of force had been breached, 'force drift' would almost certainly begin … And if left unchecked, force levels including torture, could be reached" (2009, 110).

5.9 Putting "Force Drift to Work

One factor contributing to the loss of control in "force drift" is the objectification and dehumanization of the other. Acts of coercion and brutality are more likely to occur when there is a power differential between those involved. And when the victim has a lesser moral status, as with so-called "unlawful" or "unprivileged" combatants, it's even harder to stop the slide into abuse. Visions of mob violence come to mind, in that normal inhibitions get swept aside.

Humanities Professor Colin Dayan points out the consequences; namely, "Once stigmatized categories are created, whether they are labeled 'security threat groups' or 'illegal enemy combatants,' torture can be administered readily by those in power" (2007, 57).

Here's where categorizing others as "enemies" opens the door on abuse and mistreatment. The dehumanization of our enemies can swiftly lead to disrespect and degradation. In short, the "enemy" merits few protections. With that, notions of non-maleficence and beneficence are made inoperable.

5.10 Mental Pain and Torture

Another dimension of pain involves the mental or psychological aspects. Long after the physical scars have faded and the pain has been relegated to past memories (or nightmares), there may still be long-term effects. Such effects become more apparent as time goes on.

For example, in 2016, *The New York Times* investigated the mental pain of torture of suspects who were rounded up after the 9/11 attacks. They found that psychological and emotional scars haunted the men because of the interrogations at secret C.I.A. "Black Sites" and at Guantanamo Bay. On October 21, 2016 The Editorial Board asserted that,

> A disturbingly high number of these men were innocent, or were low-level fighters who posed so little threat that they were eventually released without charge. Yet despite assurances from lawyers in the Department of Justice that "enhanced interrogation techniques" should have no negative long-term effects, *The Times* found that many of the men still suffer from paranoia, psychosis, depression and post-traumatic stress disorder related to their abuse. They have flashbacks, nightmares and debilitating panic attacks. Some cannot work, go outside, or speak to their families about what they went through.

The U.S. Senate Report shows us how bizarre this can be, in detailing the interrogation of Ridha al-Najjar (2014, 60). He was held at the CIA "Black Site" Cobalt outside of Kabul, Afghanistan, where the mental aspects of his torture were exploited.

The conditions were horrific: In the regular cells, James Wilkinson (2017) reports, they were shackled to a metal ring in the wall, and given a bucket to use as a toilet. In the sleep deprivation cells they were shackled by their hands to the ceiling, and made to defecate in diapers. When diapers were not available, they stood bare from the waist down, or defecated into makeshift diapers created using duct tape. The cells were unheated, and subjected to blaring music.

The U.S. Senate Report on Torture (2014) details al-Najjar's treatment and reveals how vicious were the conditions used to break him—which they succeeded in doing.

> The CIA discussed its interrogation strategy of al-Najjar during June and July [2002] … One cable, dated 16 July 2002, was sent to the CIA Station in Country [redacted], "suggesting possible interrogation techniques to use against Ridha al-Najjar, including: utilizing 'Najjar's fear for the well-being of his family to our benefit' … using 'vague threats' to create a 'mind virus'; that would cause al-Najjar to believe that his situation would continue to get worse … manipulating Ridha al-Najjar's environment using a hood, restraints, and music; and employing sleep deprivation through the use of round-the-clock interrogations."
>
> By 26 July 2002, CIA officers were proposing "breaking Najjar" through the use of "isolation, sound disorientation techniques, sense of time deprivation, limited light, cold temperatures, sleep deprivation" (2014, 60).

The Report goes on to say that the CIA's interrogation plan for al-Najjar included for the use of "loud music, worse food, sleep deprivation and hooding." In addition,

> al-Najjar was tortured throughout August and September 2002, and by 21 September one CIA cable was clear that he was now "clearly a broken man" and "on the verge of complete breakdown" as a result of isolation. Indeed, al-Najjar was now "willing to do whatever the CIA officer asked" (The Rendition Project).

A second example of the mental aspects of physical torture is the case of Mohamed Ben Soud, who was detained by the CIA in Afghanistan. According to a *New York Times* report by Sheri Fink and James Risen (2017), he was subjected to being locked in small boxes, slammed against a wall and doused with buckets of ice water while naked and shackled. The consequences of this abuse are significant, as the following indicates:

> He said he still suffered from nightmares, fear, mood swings and other psychological injuries as a result of his captivity. "It comes to me during my sleep and as if I'm still imprisoned in that horrible place and still shackled," he said in his deposition, through a translator. "I get the feeling of worry about my future and about the fear that this could happen again" (Fink and Risen 2017).

Self-inflicted fear and other kinds of distress, as shown with both al-Najjar and Ben Soud, can be mentally debilitating and cause significant harm. We see this with the use of psychological manipulation via the "mind virus."

5.11 Employing a "Mind Virus"

Mike Doherty (2015) had the opportunity to interview an ex-CIA interrogator. He spoke of the way mental suffering was put to use when questioning (and trying to break) a suspect. To achieve that goal he employed a "mind virus." He explains:

> We capitalize on certain psychological techniques that aren't related to personal threat. Let's say for example you walk in the office tomorrow, and one of your co-workers comes running up to you and says, "The boss wants to see you right away." ... Do you immediately think, "Today's a great day. I'm going to get promoted"? Probably that's not our thought process. The message you have received has planted what behaviorists would term a "mind virus." We don't know whether or not that would create fear or uncertainty—we don't know where that virus is going in your mind. ... It's fear that's self-inflicted (2015).

He acknowledges that this procedure is not always productive. "Mind viruses can be misused," he says, "so if you're in custody and I walk in, and all of a sudden I pull out a Glock 9 mm and set it on top of the desk, that's the mind virus, right? But that's not the way to do it. That's the wrong message. We believe that's counterproductive" (Doherty 2015).

In the case of mental aspects of pain from torture, victims may feel responsible for their suffering—their "mind virus"—as if they have some control over its effects. This can get exploited. We see this in the CIA manual recommending that the subject be manipulated into blaming themselves for their suffering.

For example, when a detainee is told, "You leave me no other choice but to ... " the victim may interpret "You leave me" as something they're in control of (Wisnewski and Emerick 2009, 111). This is quite different from being told "Comply or else", where no options are presented to them. Blaming yourself for being harmed by another can cause long-lasting post-traumatic stress. This makes it a powerful tool and quite useful in establishing control of the victim.

5.12 Making Victims Feel Responsible for Their Pain

Historian Alfred McCoy discusses how people can be manipulated into feeling responsible for their mistreatment. He says,

> Once the subject is disoriented, interrogators can then move on to the stage of self-inflicted pain through techniques such as enforced standing with arms extended… In this latter phase, victims are made to feel responsible for their own suffering, thus inducing them to alleviate their agony by capitulating to the power of their interrogators (Wisnewski and Emerick 2009, 114–115).

Offering options to detainees so they feel responsible for their own pain and stress can take absurd turns. Consider the choices offered in force-feeding. Colin Dayan observes that:

> Talking to a group of reporters about the chair to which detainees were strapped during the insertion of the feeding tubes, General John Craddock, the head of the United States Southern Command, said, "it's not like 'The Chair.' It's pretty comfortable; it's not abusive." He explained how his soldiers gave detainees a choice of colors for feeding tubes–yellow, clear, and beige—adding, "They like the yellow" (2007, 74–75).

Being able to choose the color of your straw for force-feeding is like getting to pick the belt you are going to be beaten with. In any case, by bringing the victim into the decision-making, the blame for the abuse gets spread wider. One result is this strengthens the mental component of torture.

5.13 Clustering: Combining Methods of Abuse

The Senate Report on Torture also confirms the theory of a "slippery slope" in interrogation settings, namely that torture by its very nature escalates in the severity and frequency of its use beyond the approved techniques.

This raises two key points. First, while the techniques are evaluated individually, these techniques were designed to be used in combination, thus enhancing pain and disorientation. This is a game-changer in terms of the effects on the victim of clustered techniques.

Second, to comprehend the severity of the effects of these techniques, it is essential to consider the context of their use. In terms of both long and short-term psychological effects, there is no meaningful equivalence between waterboarding when used in SERE (Survival Evasion Resistance Escape) training of soldiers who volunteered and consented to the procedure and who trusted the interrogator to protect their safety, and waterboarding a high-value detainee in a setting where the victim fears for his life (PHR 2009, 1). The two situations have little in common in terms of the psychological component.

5.14 How Does Language Figure in?

As Luis Justo (2006) asserts, we need to discuss the "philosophy of torture" (if such a concept exists). This calls for an examination of the various terms of the debate, including ones that obscure or trivialize the pain of torture.

The use of euphemisms such as "enhanced interrogation" to describe torture should be discredited, Justo recommends, since it contributes to public mystification (2006, 1463). With that mystification comes a reluctance to see how much brutality has become tolerable. Furthermore, such "public mystification" entails using or creating terms that have no ready reference to previous uses of language. If a phrase like "enhanced interrogation" can replace "torture," then concerns are less likely to be raised and there's less chance for condemnation—or public outrage.

5.15 Twisting Language

As noted in Chap. 2, the term "prisoners" has generally been replaced by "detainees" to allow fewer protections in the war on terror. "Detainees" are not considered protected by the Geneva Conventions or international ethical codes and treaties. Consequently, "detainees" lack the moral status accorded "prisoners," setting the stage for actions and policies that makes them vulnerable to mistreatment.

The low moral status of detainees relative to that of prisoners was sealed by a linguistic fiat drawing a sharp distinction between the two groups. This is borne out by the rejection of the Geneva Conventions' guideline requiring the prompt registration of prisoners. That has not been perceived as binding in the case of detainees. Basically, those who are not prisoners do not merit the same treatment; thus their vulnerability.

Once this distinction is in place, attention then could turn to what is allowable treatment. What was prohibited in interrogating "prisoners" was not seen as applicable to "detainees." Even citizenship cannot save a terrorist suspect from being swept up into the class of "detainees" and having but a tenuous moral status. Look, for instance, at the treatment of John Walker Lindh and Jose Padilla, both American citizens. The abuse they experienced indicates how easily human rights can be thrust aside.

We see this also with innocent individuals like Khaled el-Masri On Dec. 31, 2003, el-Masri was on a tourist bus headed for a vacation at the Macedonian capital, Skopje. At the border stop he was held back and hauled off the bus.

> He said that after being kidnapped by the Macedonian authorities at the border, he was turned over to officials he believed were from the United States. He said they flew him to a prison in Afghanistan, where he said he was shackled, beaten repeatedly, photographed nude, injected with drugs and questioned by interrogators about what they insisted were his ties to Al Qaeda (Van Natta Jr. and Mekhennet 2005).

All too often the haste to apprehend potential terrorists overrides any presumption of innocence until they are charged with a crime, much less proven guilty. El-Masri has not been able to seek legal remedies, because the U.S. government successfully invoked a "state secrets" privilege (Goldberg 2009).

5.16 And Then It's All Couched in Secrecy

Unfortunately, there has been a great deal of secrecy about policies that have been instituted and actions that have been taken. One of the most disturbing aspects of secrecy is that detainees lack access to the evidence against them. Clearly, it is hard to defend yourself when the charges and evidence have been withheld.

Access to a lawyer is difficult if not impossible, adding to the obstacles facing detainees. And with the fate of an indefinite detention, once a detainee can feel like *always* a detainee, given there may be no end in sight for those stuck there in this situation.

5.17 The Impact of Manipulating Language

The policies and regulations that inform medical personnel rest, to a great degree, on definitions, categories, connotations and concepts. The connotations of words can play a significant role in shaping policies. Look, for example at "rampage" v. "massacre" v. "act of terrorism", "extremist" v. "terrorist" and "rebel" v. "soldier." The use of language and reshaping of terms carries weight. Our thoughts and policies are guided by the words we use. As a result, employing terms like "detainees" instead of "prisoners" makes all the difference.

Here's where language comes into play: our soldiers are "lawful combatants," whereas "insurgents" are "unlawful combatants" and not bound by the same standards. To add fuel to the fire, the use of such terms as "illegal combatants," "foreign combatants", and "unprivileged enemy combatants" functions as a signal to distinguish the good guys from the bad guys, the enemies. This lays a foundation for the lesser rights accorded "detainees" and makes it easier to slide into abuse.

And so divisions are established. There are five categories that a terrorist suspect can fall into: (1) "Insurgents"—those yet to be captured; (2) "Prisoners"—those charged with a crime and awaiting trial; (3) "Detainees"—those held at a detention center and have neither been charged nor convicted of a crime; (4) Detainees undergoing "rendition", the transport to a country known to allow torture (they are then considered "rendered"); (5) "Ghost detainees" or "Ghost soldiers"— those held without charge at secret detention centers— "Black Sites"— far from the accessibility of the Red Cross or legal counsel. They are said to have undergone "extraordinary rendition."

Detainee attorney Joseph Margulies notes that: "The horror story of the post-9/11 world is that any foreign national anywhere in the world can be plucked from the streets of anywhere, whisked off to another country, never be heard from again and be utterly beyond the reach of the law" (as noted in Teays 2008). That is bad enough. Even worse is the public indifference to the mistreatment—including torture and indefinite detention—of thousands of innocent people caught up in this war on terror.

The dismal status of detainees explains why Executive Director of the Center on Law and Security at the New York University School of Law, Karen J. Greenberg refers to them as "this nebulous class of persons." The result, she says, is a "new category of person" that has "extricated the United States from the international obligations that have governed the treatment of prisoners in armed conflict since the middle of the nineteenth century." The resulting shift was used to redefine torture; thus allowing coercive interrogation techniques such as waterboarding (2006, xi, as noted in Teays 2011, 72).

5.18 The Language Games

All the language games need to be examined. Look, for instance, at the *Hamdi vs. Rumsfeld* decision allowing the president to detain an enemy combatant. "How do you make that determination?" wonders 4th Circuit Judge Diana Motz. "When I call someone an ostrich, I look in the dictionary for a definition. But what did the president look to in determining whether he [Yasser Hamdi] was an enemy combatant?" (as noted by Dahlia Lithwick 2007). Defining such terms is long overdue, as is their free-wheeling application.

Judge Motz isn't the only one questioning the use and misuse of language in order to put desired policies in place. Social Anthropologist Tobias Kelly also discusses the language of torture and the difficulties in nailing down the concept. Those difficulties are exacerbated by the lack of transparency regarding policies and actions. "The recognition of torture presents unique challenges," Kelly argues.

> Torture's particular stigma, as one of the most universally recognized violations of human rights, raises the stakes for those states accused of torture. Very few, if any, states willingly admit that they participate in torture. Furthermore, despite its apparent moral absolutism, torture remains a notoriously slippery category to define because its meaning constantly shifts under pressure … Any attempt to recognize torture must therefore overcome serious political, legal, and epistemological hurdles (2009, 778).

Given all the hurdles around torture, one of the epistemological tasks is to pay heed to the power of labeling, particularly when it is used to dehumanize people, as we find with prisoners and detainees. That power should not be underestimated.

5.19 Strategies for Justifying "Forceful"/"Enhanced" Interrogation

It is instructive to get a sense of techniques of abuse that were sanctified at the highest levels. See, for example, the "torture memos" of the Bush administration, the first of which was drafted by John Yoo and supported by White House counsel Alberto Gonzales. Andrew Cohen. of *The Atlantic* reports that the memo was sent to "all the key players of the Bush administration" (2012). Evidently their assumption that the infliction of physical or mental pain has its merits prevented them from declaring such practices off-limits.

In order to justify the use of forceful interrogation techniques, the U.S. Defense Department adopted two strategies to get around the prohibition on torture, notes Werner G. K. Stritzke (2009). The first asserts that the president's authority to manage military operations is uninhibited by international law and that the use of torture in an interrogation is an act of national self-defense and thus may not be violating the prohibition (2009, 31). This effectively disavows the role and power of international law.

The second strategy advocates for a narrow interpretation of what counts as torture. For example, a Defense Department memo argued that the administration of drugs to detainees violates the prohibition on torture *only if* intended to produce "an extreme effect." Similarly, then U.S. Assistant Attorney-General Jay Bybee insisted that, "Torture must be equivalent in intensity to the pain accompanying serious physical injury, such as organ failure, impairment of bodily function, or even death" (Stritzke 2009, 31).

5.20 What Are the Consequences of Such Strategies?

In setting down such restrictions, Bybee effectively narrowed the domain of what counts as torture, so only the most egregious acts would qualify. Given the stakes are high with regard to human rights, it should not be surprising that Bybee would be averse to having the term "degrading treatment" applied to an action sanctioned on the part of the government.

Bybee recommended an interpretation of torture emphasizing the *intention* of the interrogator. An individual could only be said to commit torture if they *intended* to do so. "Thus any interrogator who tortured but later claimed that his intention was to gain information rather than inflict pain was not guilty of torture" (Wisnewski and Emerick, 114–115). That means someone could only be thought guilty of torture if they *sought* to harm or maim the other. Therefore, "Threatening death and inflicting pain would be actionable only when the interrogator intended to harm" (Dayan 2007, 68–69). This puts intentions—not actions—at center stage.

Think about this interpretation. Redefining the term shrinks its range of applicability, or so it is claimed. Wisnewski and Emerick note the consequences:

> This new interpretation of torture removes so much moral substance from the term ... [so only] the category of "dehumanizing, torture" would remain ... As a result, the Geneva Conventions were interpreted less and less broadly, and thus their power to guide interrogations was weakened (2009, 114-15).

Adopting such an approach would have significant repercussions. It would shift culpability from the perpetrators' actions to their mental state. So long as someone wasn't *contemplating* harm they should not be held responsible for their actions. Only. If you are determined to—planning to—inflict pain or suffering should you be thought committing torture. Actions should not be labeled "torture" otherwise.

Such a policy would inject a *mens rea* criteria into assessing an act we might otherwise consider torture. The question is how much emphasis should be placed on the state of mind of the perpetrator—or interrogator.

5.21 Shaping Public Consciousness

Richard Jackson (2007) asserts that officials create morality-defining narratives to shape the public's acceptance of torture. This is accomplished by replacing existing social reality that prohibits torture with a torture-sustaining reality, one resting on new morality-defining narratives. He contends that torture would then be enabled and practiced routinely.

If that were the case, not only would torturers have to be trained, but the greater society would have to be prepared and, "in a sense, trained to accept that such things go on" (Jackson 2007, 359). A series of power narratives and representations would then be endlessly reproduced until they become accepted as legitimate forms of knowledge and practice (359). Torture then becomes part of our cultural mindset and is then normalized. At that point a moral transformation will have taken place.

Given the power of narratives, we should not overlook the role of language and the media in making torture more palatable. This merits attention, since we are a society in which stories – narratives – have great appeal. We love stories of heroes vanquishing villains, battles quickly won, and enemies defeated. The dehumanization and objectification of "villains" makes their capture and torture less disturbing than if we did not employ such polar extremes in our descriptions. To make those stories accomplish the goal, the use of terms like "unlawful enemy combatants" and "cowards" come into play.

Similarly, the practice of putting hoods on prisoners or blackened goggles, masks and ear covers during transit, makes the enemy 'faceless,' dehumanized (Jackson 2007, 362). It is much easier to inflict pain on the hooded, faceless Other than to look them in the eye and see their expression (and suffering) as they are being brutalized.

Jackson also discusses the power of images and considers the photographs of detainees (as, for instance, from Abu Ghraib) in huge piles of naked bodies as the ultimate indication of degradation. For a moment in time, he says, the 'terrorists' ceased to be individuals and their humanity dissolved (2007, 363–64). And so they are at the mercy of their captors.

5.22 The Use of Language: Medical Personnel

Narratives, images, and revised or invented terminology don't just shape the way we think of detainees. The effects have a greater sweep, extending to health professionals as well. Instead of "doctors" or "medical caregivers" we have the militarized term "medically-trained interrogators," as Philosopher Fritz Allhoff calls them. In his view, "The interrogator's primary task is to facilitate the acquisition of information, not to heal" (2008, 101).

Allhoff is not the only one reshaping concepts. Some perceive doctors in interrogations as *combatants* to whom the Hippocratic Oath does not apply (Stephens 2005). This fundamentally changes what would count as doctors' fiduciary duties and dismantle the patient-doctor relationship.

5.23 Conclusion

We see how this gets played out: Bioethicists M. Gregg Bloche and Jonathan H. Marks (2005) assert that evidence reveals that medical personnel shared confidential documents with potential interrogators. In addition, physicians helped design interrogation strategies, including sleep deprivation and other coercive methods tailored to detainees' medical conditions. Medical personnel also coached interrogators on questioning techniques, Fritz Allhoff points out (2006, 393).

A medical degree is not a sacramental vow, contends David Tornberg, Deputy Assistant Secretary of Defense For Health Affairs. "It is a certification of skill." As he sees it, a medical degree becomes a practical diploma carrying no more ethical weight than a plumber's (Koch, 2006, 250).

Not all agree with Tornberg's downgrading of the medical degree to strip it of its moral significance and ignore the professional codes intended as guidance. See, for example, the UN Committee Against Torture (United Nations 2009). However, many recognize the conflicts that can arise when health professionals are put in a key role with patients. This is acknowledged in the Baha Mousa Public Inquiry Report (2011) regarding the brutal death of an Iraqi civilian held by the British Army:

> The report recognizes the problem of "dual loyalty" in which doctors, particularly in conflict situations, are often torn between their patient and their employer. It also identifies legal and military systems that force doctors into situations where they may risk their jobs or put their own lives in danger by refusing to participate or report torture that they have witnessed (2011).

Journalist Sophie Arie also notes the concern that British doctors working at immigrant detention centers fail to examine properly or report injuries asylum seekers have received before arriving in the UK.

> This can lead not only to their not receiving treatment they need but also to a lack of evidence supporting the case for asylum for those who have been tortured or abused elsewhere. Military doctors have also failed to report signs of abuse in detainees (Arie 2011).

The United States faces the same issues regarding the complicity of doctors in abusing detainees, hiding it or failing to report it. The 2004 CIA Inspector General's report confirms that health professionals were involved at every stage in the development, implementation and legitimization of the torture program.

Legal Officer Stephanie Erin Brewer and Social Psychologist Jean Maria Arrigo (2009) point out that, "Some commanders may simply order medical personnel to place their loyalty to their country over their medical care for a detainee, including trading medical treatment for information during interrogations" (2009, 11).

This is a problem in need of a solution. Health professionals need to meet this head on and become advocates for change—change rooted in human rights.

Works Cited

Adams, Guy B., Danny L. Balfour and George E. Reed. 2006, September–October. Abu Ghraib, Administrative Evil, and Moral Inversion: The Value of 'Putting Cruelty First.' *Public Administration Review* 66 (5): 680–693. http://www.jstor.org/stable/3843897. Retrieved 2 January 2018.

Agin, Dan. 2009, April 22. How It Is: Psychiatrists, Physicians, and Torture. *Huffington Post*. https://www.cchrint.org/2009/09/08/the-huffington-post-how-it-is-psychiatrists-physicians-and-torture/. Retrieved 2 January 2018.

Allhoff, Fritz. 2006. Physician Involvement in Hostile Interrogations. *Cambridge Quarterly of Healthcare Ethics* 15: 392–402. http://files.allhoff.org/research/Physician_Involvement_in_Hostile_Interrogations.pdf. Retrieved 1 May 2018.

———. 2008. *Physicians at War: The Dual-Loyalties Challenge*. New York: Springer.

Arie, Sophie. 2011, September 9. Doctors Need Better Training to Recognise And Report Torture. *British Medical Journal* 343. http://www.bmj.com/content/343/bmj.d5766. Retrieved 10 June 2018.

Baha Mousa. 2011.The Report of the Baha Mousa Inquiry. www.bahamousainquiry.org/report/index.htm. Retrieved 10 June 2018.

Bloche, M. Gregg and Jonathan H. Marks. 2005, January 6. When Doctors Go to War. *New England Journal of Medicine* 352 (1): 3–6. https://www.nejm.org/doi/full/10.1056/NEJM200504073521423. Retrieved 10 June 2018.

Brewer, Stephanie Erin, and Jean Maria Arrigo. 2009. Preliminary Observations Why Health Professionals Fail to Stop Torture in Overseas Counterterrorism Operations. In *Interrogations, Forced Feedings, and the Role Of Health Professionals*, ed. Ryan Goodman and Mindy Jane Roseman. Cambridge, MA: Human Rights Program, Harvard Law School.

British Broadcasting Corporation. 2011, September 8. Baha Mousa Inquiry: 'Serious Discipline Breach' by Army. bbc.com. http://www.bbc.com/news/uk-14825889. Retrieved 10 June 2018.

Cohen, Andrew. 2012, February 6. The Torture Memos, 10 Years Later. *The Atlantic*. https://www.theatlantic.com/national/archive/2012/02/the-torture-memos-10-years-later/252439/. Retrieved 30 March 2019.

Dayan, Colin. 2007. *The Story of Cruel and Unusual*. Cambridge, MA: The MIT Press.

Doherty, Mike. 2015, April 23. A Former CIA Officer Explains How to Apply Interrogation Techniques to Everyday Life. *Vice*. https://www.vice.com/en_us/article/4wbygn/three-former-cia-officers-want-to-help-you-apply-interrogation-techniques-to-everyday-life-113. Retrieved 10 June 2018.

Editorial Board, *The New York Times*. 2016, October 21. Torture and Its Psychological Aftermath. *The New York Times*. https://www.nytimes.com/2016/10/21/opinion/torture-and-its-psychological-aftermath.html. Retrieved 10 June 2018.

Fink, Sheri and James Risen. 2017, June 20. Psychologists Open a Window on Brutal CIA Interrogations. *The New York Times*. https://www.nytimes.com/interactive/2017/06/20/us/cia-torture.html. Retrieved 10 June 2018.

Ford, Matt. 2014, December 13. Antonin Scalia's Case for Torture. *The Atlantic*. https://www.theatlantic.com/politics/archive/2014/12/antonin-scalias-case-for-torture/383730/. Retrieved 10 June 2018.

Goldberg, Nicholas. 2009, February 15. 'State Secrets' on Trial. *The Los Angeles Times*. https://www.latimes.com/la-oe-secrets15-2009feb15-story.html. Retrieved 10 June 2018.

Greenberg, Karen J. 2006. *The Torture Debate in America*. Cambridge, MA: Cambridge University Press.

Human Rights Watch. 2000, October, Welcome to Hell: Arbitrary Detention, Torture, and Extortion in Chechnya - The Chernokozovo Detention Center. *Human Rights Watch Report*. https://www.hrw.org/reports/2000/russia_chechnya4/detention-center.htm. Retrieved 10 June 2018.

Ignatius, David. 2005, December 16. Stepping Back From Torture. *The Washington Post*. http://www.washingtonpost.com/wp-dyn/content/article/2005/12/15/AR2005121501438_Comments.html. Retrieved 10 June 2018.

Jackson, Richard. 2007, July. Language, Policy and the Construction of a Torture Culture in the War on Terrorism. *Review of International Studies* 33 (3): 353–371. https://www.cambridge.org/core/journals/review-of-international-studies/article/language-policy-and-the-construction-of-a-torture-culture-in-the-war-on-terrorism/5006D871A4E5550BE5EBA0F10CCEDACF. Retrieved 10 June 2018.

Justo, Luis. 2006, June 24. Doctors, Interrogation, and Torture. *British Medical Journal* 332 (7556): 1462–1463. http://www.jstor.org/stable/25689660. Retrieved 10 June 2018.

Koch, Tom. 2006, May. Weaponizing Medicine. *Journal of Medical Ethics* 32 (5): 249–252. https://www.jstor.org/stable/27719620?seq=1#page_scan_tab_contents. Retrieved 10 June 2018.

Lithwick, Dahlia. 2007, February 1. The Third Man: The 4th Circuit Does One More Round on Enemy Combatants. *Slate*. http://www.slate.com/articles/news_and_politics/jurisprudence/2007/02/the_third_man.html. Retrieved 10 June 2018.

Maio, Giovanni. 2001, May 19. History of Medical Involvement in Torture—Then and Now. *Department of Medical History The Lancet* 357 (9268): 1609–1611. https://www.thelancet.com/journals/lancet/issue/vol357no9268/PIIS0140-6736(00)X0242-5. Retrieved 10 June 2018.

Nel Noddings. 1989. *Women and Evil*. Berkeley, LA: University of California Press.

Physicians for Human Rights. 2009, August. Aiding Torture: Health Professionals' Ethics and Human Rights Violations Revealed in the May 2004 CIA Inspector General's Report.

Savage, Charlie. 2004, December 27. CIA Resists Request for Abuse Data. *Boston Globe*. http://www.boston.com/news/globe/. Retrieved 10 June 2018.

Stephens, Joe. 2005, January 6. Army Doctors Implicated in Abuse. *The Washington Post*. https://www.washingtonpost.com/archive/politics/2005/01/06/army-doctors-implicated-in-abuse/f44fe582-81d5-4eba-80da-a0edd78bbc61/?utm_term=.da01bb242159. Retrieved 10 June 2018.

Stritzke, Werner G.K., Ed. 2009. *Terrorism and Torture: An Interdisciplinary Perspective*. Cambridge: Cambridge University Press.

Teays, Wanda. 2008. Torture and Public Health. In *International Public Health Policy and Ethics*, ed. Michael Boylan. Dordrecht: Springer.

———. 2011. The Ethics of Otherness. In *Morality and Global Justice Reader*, ed. Michael Boylan. Boulder, CO: Westview Press.

The Rendition Project. Ridha al-Najjar. https://www.therenditionproject.org.uk/prisoners/najjar.html. Retrieved 10 June 2018.

The U.S. Senate Committee. 2014, December 8. *The Senate Committee Report on Torture*. https://www.nytimes.com/interactive/2014/12/09/world/cia-torture-report-document.html. Retrieved 10 June 2018.

United Nations. 2009. The UN Committee Against Torture: Human Rights Monitoring and the Legal Recognition of Cruelty. *Human Rights Quarterly*. https://www.jstor.org/stable/40389967. Retrieved 10 June 2018.

Van Natta Jr., Don and Souad Mekhennet. 2005, January 9. German's Claim of Kidnapping Brings Investigation of U.S. Link. *The New York Times*. http://www.nytimes.com/2005/01/09/world/europe/germans-claim-of-kidnapping-brings-investigation-of-us-link.html. Retrieved 10 June 2018.

Wilkinson, James. 2017, October 10. CIA Black Site 'Cobalt'. *The Daily Mail (UK)*. http://www.dailymail.co.uk/news/article-4965130/Grim-facts-emerge-CIA-black-site-Cobalt.html#ixzz58S4T8vjo. Retrieved 10 June 2018.

Wisnewski, J. Jeremy, and R.D. Emerick. 2009. *The Ethics of Torture*. London: Continuum.

Chapter 6
Degradation and Dehumanization

> *You just had to tune it out. You couldn't let it get to you. You got numb. But it catches up to you later, when you get home. Like, I slept fine while I was there, but now I have nightmares.*
> —Army Reservist Joseph M. Darby.

Overview

In Part II we look at the boundaries of torture and the ethical issues they bring to light. This chapter focuses on the helplessness of the victims subjected to degradation and dehumanization. As we see, torture is not just about inflicting pain; it is also about asserting power. Treating others in demeaning ways is an effective vehicle for accomplishing that goal.

As the examples testify, the injury can be long- lasting. This can be seen in five practices that stand out on the degradation and dehumanization scale. Together they paint a picture of human cruelty, often leaving the victims damaged, and defeated. These are hooding, forced nudity, force-feeding, waterboarding, and humiliation. We will look at each one, some of which are combined, clustered, to magnify their effect.

- Hooding and forced nudity both strip detainees of their identity, making their abuse much easier than if we could see their faces and interact with them. By putting a bag or two over their heads, we shield ourselves from their suffering. Both tactics diminish their sense of self. Furthermore, as Major General George Fay observed, forced nudity can contribute to an increase in sexual assaults.
- Force-feeding is also degrading and dehumanizing. It is generally a response to a hunger strike. That it goes against the right to self-determination is examined here. Doctors and nurses are involved in force-feeding detainees—a controversy that the profession has responded to.
- Waterboarding, alias "simulated drowning," also underscores the helplessness of the victim. We get a sense of its long history and the role it has played in the interrogation of high-profile suspects. That it is terrifying is born out by the

testimony of four Americans who were waterboarded as part of SERE training. However, that did not stop it from being used on high-profile detainees.
- The fifth practice we examine is the use of humiliation. Three techniques are especially debilitating—forced shaving, prolonged diapering, and sexual assault.

6.1 Introduction

The Economist calls it *a journey into hell:* On August 28, 2012 six members of the local Human Rights Council went to Romeu Gonçalves de Abrantes prison in João Pessoa, Brazil. They found filthy, overcrowded cells holding sick, thirsty prisoners, some with untreated injuries. The locked punishment wing reeked of vomit and feces. A camera stuck in a ventilation shaft revealed images of naked prisoners crowded into bare, unlit cells.

Lawyer Marcos Fuchs said he has seen "cells built for eight men holding 48, cases of gangrene and tuberculosis left untreated and prisoners kept in unventilated metal shipping containers under the baking sun." A 2009 congressional inquiry documented routine beatings and torture by guards and prisoners locked up without daylight for months (*The Economist* 2012).

A second journey into hell, that of "American Taliban" John Walker Lindh:

> The Marines forcibly stripped him naked, using scissors to cut off his clothing. He was bound, naked, with duct tape to a gurney, his hands tied together with excruciatingly painful plastic straps. It was December [in Afghanistan] and freezing cold. The gurney, with John strapped onto it, was placed by the Marines in an unheated metal shipping container. He trembled uncontrollably from the cold. He cried out in pain.
>
> An AK-47 bullet and shrapnel shards were left, untreated, in his lower extremities. A military doctor who later examined him said John's wounds were left festering and untreated. The Marines taunted him and repeatedly threatened to kill him. He was kept in this condition for two days and two nights. He was then removed from the shipping container ... His wounds were still untreated (Lindh 2014).

And *a third journey into hell,* that of Irish Republican Army (IRA) detainee Seán McKenna: McKenna was held with 13 others by the British army in 1971 and subjected to what *The Irish Times* (2016) called "horrific interrogation from which they never recovered." After a few days of physical abuse and humiliation, McKenna says he was hooded, tortured, and interrogated. He describes the experience:

> My head was spinning ... my mind went wild, I was crying, I couldn't stand up and I was trying to grip the wall ... I couldn't even remember my name or my children's names ... My head still won't focus right ... I detest being alone ... I never imagined anyone could be so cruel to his fellow man ... I don't think I will ever be the same again" (*The Irish Times* 2016).

Each of these three "journeys into hell" illustrate the depravity that can be inflicted on suspects, detainees, prisoners. Moreover, it takes place around the globe and is not a strategy of one faction, one group, or one nation.

Torture is not just about inflicting pain, though that is the norm. It is also about asserting power. The victim is helpless, unable to assert control over what's happening. The effects can be debilitating and long reaching, causing lasting harm, as McKenna indicated when he said, "I don't think I will ever be the same again."

In this chapter we will examine five tactics that stand out on the degradation and dehumanization scale. Together they paint a picture of human cruelty, often leaving the victims damaged, and defeated. These are hooding, forced nudity, force-feeding, waterboarding, and humiliation. We will look at each one, some of which are combined—clustered—to enhance their effect.

Victims are marked by the cruelty. For example, lawyer-activist Jennifer K. Harbury remarks that, "All of the Latin American torture survivors remembered the hoods, the constant beatings, the nakedness, the electrodes, and the rapes and sexual humiliations as well" (Harbury 2005, 14). The combination of brutal and degrading treatment is pretty much guaranteed to have lasting consequences.

Let's start with the use of hoods.

6.2 First Strategy: Hooding

Putting a hood or sandbag over people's heads is a quick way to dehumanize and terrify them. As Richard Jackson says, they are now faceless (2007, 362). The consequence is that the primary means of relating to each other has been severed. They can be seen but cannot see who or what they are interacting with, putting them at a considerable disadvantage. They are stripped of their identity—their individuality effectively erased with their heads in bags. They are then at the mercy of those in control, something the perpetrators use to their advantage.

Physicians for Human Rights (PHR) sees the purpose of hooding as creating sensory deprivation, isolation, and confusion (2009). That is a key function that hoods serve. They disorient the captives and put them in a state of fear as to what fate awaits them—a fate they cannot prepare for.

The International Red Cross points out an additional use of hooding beyond instilling a sense of fear and disorientation; that is a dependency on their captors (NPR 2004). If people are unable to see (and with ear muffs unable to hear), they cannot move about freely or easily function in their environment. In addition,

> [Hooding is also] to prevent them from breathing freely. Sometimes two bags are used. In addition, hooding was sometimes used in conjunction with beatings thus increasing the anxiety has to when the blows would come. It also allowed the interrogators to remain anonymous and thus to act with impunity (NPR 2004).

A second function of hoods is to ensure anonymity for the guards and interrogators, making it easier to assert dominance. They can then exercise power and aggression without the distraction of empathy. By turning detainees into faceless objects, guards are less likely to care for them or show restraint under pressure. The not-quite-human Other is much less likely to elicit sympathy than an identifiable person who can convey misery and suffering. Even something as simple as a conversation is unlikely in such a scenario. The verbal interchange would probably then consist of the one with the power barking commands at the one who is powerless.

This may seem convenient for interrogations or for exercising control, but is at a cost for all concerned. You can't treat others in a degrading manner without harming them. And you can't treat others in a degrading manner without harming your own moral fiber. Eventually there will be repercussions.

6.3 The Cost Is Not Insignificant

The reality is that it's much easier to care about others if we are on the same level and moral status. A caring disposition is also created when we have assumed fiduciary duties to benefit or protect the other. This requires us to acknowledge their individual dignity. Hooding, however, locks in place a power differential between the guard—or medical personnel—and the detainee that can easily turn adversarial. It's no wonder that things can escalate and the door to abuse and mistreatment swings wide open.

It's hard to imagine a variation of hooding that's more barbaric, but duct tape may take the prize. The FBI reported investigators using duct tape it wrap a detainee's head because he wouldn't stop quoting the Koran. A Pentagon review in March 2008 did not condemn that strategy: The forcible use of duct tape to silence Ali al-Marri from chanting was found "acceptable" (Mazzetti and Shane 2008).

Where is the voice of moral revulsion? Where is the public outcry? Do we have no shame? Fortunately, not all stay silent—the World Medical Association Declaration on the Rights of the Patient states that the use of hoods during a medical evaluation is absolutely unacceptable. As a result, health professionals should object to the use of hoods in treating patients and speak out against this demeaning practice.

A second example of dehumanization of detainees—also widely used—is forced nudity.

6.4 Second Strategy: Forced Nudity

Forced nudity in prisons and detention centers is on a similar level to hooding in terms of objectification and dehumanization. Stripping people of clothing and making them wear a bag or two over their head (or blindfolds and ear muffs) reinforces the fact that the ones with clothes have power over those without. To be exposed and disoriented by such an egregious form of depersonalization leaves a great deal to be desired, unless there is good reason to employ such tactics. Only rarely, if at all, should forced nudity be an option and then only for a limited time.

That this practice became systemic indicates the depravity functioning throughout the chain of command. It is not the act of a handful of wrongdoers. Not in the slightest. It was planned and approved to be a legitimate means of handling prisoners and detainees, functioning as a control measure and a way to enhance interrogations.

In some places it became the norm. For instance, *The New York Times* reports that, "forced nudity of prisoners was pervasive in the military intelligence unit of Abu Ghraib, so much so that soldiers later said they had not seen 'the whole nudity thing,' as one captain called it, as abusive or out of the ordinary" (Zernike and Rohde 2004). Given the extent of its use, forced nudity does not seem out of the ordinary. That says everything.

In the Fay Report (2004), Major General George Fay spoke disapprovingly of the use of forced nudity and suggested it would open the door to other kinds of abuse. He notes that forced nudity has not historically been included in interrogations, but was thought to be an "effective technique for which no specific written legal prohibition existed." With no prohibitions in place it was "employed routinely and with the belief it was not abuse". Nevertheless, it contributed to an escalating dehumanization of the detainees (Fay 2004, 122). Certainly rape, sexual abuse, taunting, and humiliation becomes easier to commit when the person is naked, defenseless, and vulnerable.

Also demeaning is the rampant use of photography. Photographing nude detainees has been done extensively to assert power, humiliate, and diminish them. Richard Jackson (2007) speaks of its harms: "They ceased to be individuals and their humanity dissolved," he observes. In his view, the photographs of prisoners piled on top of each other mirrors images of naked corpses in the Holocaust's concentration camps. This was enabled, he argues, by a public discourse that defines 'other' as inhuman 'animals' and 'parasites' (2007, 363–64). Prejudice adds fuel to the fire of mistreatment.

This much is true: The use of nude photography is a degrading practice that should not be treated lightly or dismissed as a form of locker room entertainment. In fact,

> Some human rights campaigners described the act of naked photography on unwilling detainees as a potential war crime .... In some of the photos, which remain classified, CIA captives are blindfolded, bound and show visible bruises .... "Is the naked photography a form of sexual assault? Yes. It's a form of sexual humiliation," said Dr Vincent Iacopino, the medical director of Physicians for Human Rights [and torture expert] (*The Guardian (UK)* 2016).

Iacopino reinforces the view that the use of nude photographs raises grave concerns: "It's cruel, inhumane and degrading treatment at a minimum and may constitute torture." Note also that, "International human rights law, to include the Geneva Conventions, forbids photographing prisoners except in extremely limited circumstances related to their detention, to include anything that might compromise their dignity" (*The Guardian (UK)* 2016).

In addition, the widespread use of photographs and videotapes to embarrass, threaten, and humiliate adds to the psychological stress. Human Rights lawyer Fionnuala Ní Aoláin (2016) contends that such photographs highlight the depravity and inhumanity of such detention practices. She argues that, "The existence of such photographs underscores the ways in which pornographic vignettes continue to be revealed in the detention regime, and provokes deep unease about the motivations, rewards, and culture that pervaded the detention system" (2016). She focused her remarks on the CIA, but we can generalize to other detention centers and prisons that use forced nudity. Aolain points out that,

> While the ad hoc international tribunals have never held per se that photographing a prisoner naked is torture or cruel, inhuman, and degrading treatment there is a solid jurisprudence on the harm of forced nudity and explicit recognition that this practice is a breach of both human rights and humanitarian law standards ....
>
> At the International Criminal Tribunal for the former Yugoslavia (ICTY) for example, the Kunarac Court had a two-pronged test for outrages upon personal dignity:

1. Intentionally committing an act which would generally considered to cause serious humiliation, degradation, or otherwise be a serious attack on human dignity, and
2. Knowing that the act could have that intended effect (2016).

Aolain reminds us that, "States are no longer free to enable, tolerate, or acquiesce in practices that violate the human and sexual dignity of detainees." As such, "human rights compliant states are required to address allegations of ill-treatment and sexual coercion" (2016). It is clear that duty has not been taken seriously.

Another thing we need to recognize is that such demeaning treatment is a way to instill hate on the part of detainees and contribute to further misconduct on the part of the captors and medical personnel. Both hooding and forced nudity have a detrimental effect on individual dignity and social standing in their community. As such, it could be regarded as a cultural harm. That these techniques have been used widely—globally—calls for individuals and organizations, as well as public leaders, to work for change.

Our next example of degrading treatment also brings up human rights concerns. That is force-feeding.

6.5 Three: Force-Feeding

Force-feeding is typically employed on those undergoing a hunger strike. It's use is to prevent starvation, and death. That it is controversial goes without saying: If competent adults have patient autonomy, don't they have the right to refuse treatment?

The question is whether they have the right to refuse *life-sustaining* treatment. It's one thing to reject one course of treatment and pursue another option; as, for example, radiation instead of chemotherapy to fight cancer. However, refusing all but palliative treatment is another thing altogether. It presents a moral hurdle for doctors, whose duties of beneficence and non-maleficence should be in the forefront.

Let's not forget that the very process of force-feeding can be quite painful. The restraint chairs immobilize the legs, arms, shoulders and head, causing discomfort and anxiety. Detainee lawyer Joshua Colangelo-Bryan said some strikers quit protesting after being strapped in the chairs and having feeding tubes inserted and removed so violently that they bled or fainted (Democratic Underground 2009). Given that situation, medical personnel assisting in force-feeding raises its own set of questions.

Only in the face of interminable suffering should doctors help facilitate death—if then. But when an otherwise healthy person goes on a hunger strike, should doctors intervene to stop it and put the patient back on the road to health?

> The health professionals called to intervene in a hunger strike are faced with a dilemma: commit themselves to good order and discipline or comply with best practices for providing healthcare. Handling cases of hunger strikers confronts practitioners with the ethics dilemma of managing apparent intentional behavior that carries serious morbidity or mortality, but recognizing that hunger striking is a military and political tactic, and not a medical condition (Xenakis 2017).

One of the ways hunger strikes are seen as a "military and political tactic" is that it is considered a form of "asymmetrical warfare." We don't strike back at hunger strikers by initiating our own hunger strikes, as with reciprocal acts in the battlefield. Think of air strikes in retaliation for the infliction of casualties; e.g., with bombs and IEDs. We strike back at hunger strikes with force-feeding.

6.6 Breaking Hunger Strikes with Force-Feeding

Hunger strikes foster bad publicity and may elicit sympathy from the general public. Subsequently, a group of detainees going on hunger strikes is most undesirable. The fear is that one or more will die of starvation and then become a martyr, not unlike Buddhist monks who set themselves on fire in political protests against the Vietnam War.

As a result, force-feeding has been employed, even though it violates the right of a competent adult to refuse medical treatment. That right of self-determination has been set aside when it comes to force-feeding or, even more bizarre, "rectal hydration." Monique Crawford points out that,

> Some of the unnecessary medical practices performed included "rectal feeding," known as a nutrient enema, and "rectal rehydration," both of which have been classified as procedures of "medical participation in torture" by Dr. Steven Miles, a professor of medicine … and board member of the Center for Victims of Torture.
> "You can't feed somebody this way. And so, for the U.S. government to claim that this is some sort of feeding technique, that's just totally bizarre," Dr. Miles added (Crawford 2016).

6.7 "Rectal Hydration"

Basically "rectal hydration" adds another layer of degradation beyond force-feeding. Carol Rosenberg (2019) of The *New York Times* reported on trials alleging torture at Guantánamo Bay. She cites detainee Majid Khan's lawyer who spoke of Khan being subjected to forced rectal hydration. He said, "In his second year of C.I.A. detention, according to a cable cited in the study, the agency "infused" a purée of pasta, sauce, nuts, raisins and hummus up Mr. Khan's rectum, because he went on a hunger strike." Rosenberg points out, "The C.I.A. calls this 'rectal feeding.' Defense lawyers call it rape" (2019).

That forced rectal feeding is a form of rape is not a claim that is easily dismissed. The nonconsensual violation of one's body through anal penetration by an object (with or without an intent to nourish the victim) raises legal questions about rape. The updated (2012) definition of rape set out by the US Department of Justice is: "The penetration, no matter how slight, of the vagina or anus with any body part or object, or oral penetration by a sex organ of another person, without the consent of the victim." According to the Department of Justice:

> For the first time ever, the new definition includes any gender of victim and perpetrator, not just women being raped by men. "The penetration, no matter how slight, of the vagina or anus with any body part or object, or oral penetration by a sex organ of another person, without the consent of the victim." (US Department of Justice 2012)

That any health caregivers play any role in this tactic warrants the concern of the medical profession. Such practices have justifiably called for investigation and possible sanctions.

6.8 Responding to Hunger Strikes

With regard to the physician's duties of beneficence and non-maleficence in treating a hunger striker, the World Medical Association (WMA) guidelines state, "Benefit includes respecting individuals' wishes … [and] avoiding harm means not only minimizing damage to health but also not forcing treatment upon competent people" (Crosby et al. 2007). We should also recognize that,

> In the 2006 update of the Declaration of Malta, the WMA is unequivocal: "Forced feeding contrary to an informed and voluntary refusal is unjustified … Forcible feeding is never ethically acceptable. Even if intended to benefit, feeding accompanied by threats, coercion, force or use of physical restraints is a form of inhuman and degrading treatment" (Crosby et al. 2007).

Another issue made clear by the Declaration of Malta is this: "Doctors or other health care personnel may not apply undue pressure of any sort on the hunger striker to suspend the strike. Treatment or care of the hunger striker must not be conditional upon him suspending his hunger strike" (Legislation on Line 1992).

We might ask why anyone would go on a hunger strike and put oneself at risk of death. What could possibly motivate them? Bioethicists Michael A. Grodin and George J. Annas (2006) contend that hunger strikers don't really want to die; their desire is to address issues regarding their treatment and the conditions of their confinement. They recommend that procedures should be brought into conformance with international human rights law, "not on developing novel coercive techniques to break the hunger strikes" (2006).

Grodin asserts, "This is more than just about hunger strikes — it's about the rights of competent prisoners to refuse any treatment, including refusal of food as a form of protest or demand, and about the medical ethics of physicians who are called upon to order force-feeding" (Boston University Medical Campus 2010).

One reason to go on a hunger strike and, thus, risk dying is the lack of hope and dearth of options to express despair. Death may seem preferable to what may very well be endless incarceration, the sheer cruelty of indefinite detention.

Fortunately, the response to force-feeding has galvanized at least some of the medical community. As *The Guardian (UK)* reports:

> More than 260 doctors from around the world have called on the US to stop force-feeding hunger strikers at Guantánamo Bay. They say international agreements prevent doctors from force-feeding if individuals have made an informed choice about their protest. The restraint chairs used to hold inmates while feeding tubes are inserted [...] are also banned, they say (2006).

The World Medical Association (WMA) and the AMA prohibit force-feeding. The WMA states that, "Where a prisoner refuses nourishment and is considered by the doctor as capable of forming an unimpaired and rational judgment concerning the consequences of such voluntary refusal of nourishment, he or she shall not be fed artificially"(Khamsi 2006).

The AMA further asserts that, "The decision as to the capacity of the prisoner to form such a judgment should be confirmed by at least one other independent physician. The consequences of the refusal of nourishment shall be explained by the physician to the prisoner" (Boyd 2015). Those who are mentally competent are normally seen to have the right to refuse force-feeding, even if they are prisoners and even if death can result.

6.9 What About Other Kinds of Forced Treatment?

Force-feeding and rectal hydration are not the only forced medical treatment: Human Rights Watch cites a number of abuses that takes place under medical supervision, such as forcible anal and vaginal exams and the failure to provide life-saving abortion, palliative care, and treatment for drug dependency" (Amon 2010).

Detainees also have said to be subjected to nonconsensual medical treatment and the injection of unknown (to the victim) substances. Doctors need to take a stand against all forms of cruel or degrading treatment forced upon the patient.

6.10 Not All Enable Degrading Practices

Physicians for Human Rights reports that military physicians have force-fed some of the hunger strikers at Guantánamo and they may have been pressured to do so. PHR has called for an independent group of physicians to investigate the situation and ensure there is no compulsion of health personnel to engage in force-feeding. Physicians who violate their ethical duties should face sanctions and those who resist coercion to force-feed competent detainees should be supported, as was done with the US Navy nurse who got an award for refusing to force-feed a detainee.

Physicians for Human Rights commends the U.S. Navy's decision regarding the nurse who refused to participate in the force-feeding of Guantánamo detainees. The Navy decided not to discharge him. "This decision is an important step in recognizing the right of military health professionals to recuse themselves from unethical medical practices," asserts PHR medical director Vincent Iacopino (2015).

The nurse's action had further consequences: He received an Ethics award from the American Nurses Association. Iacopino lauded this decision, saying, "Bestowing a prestigious ethics award on the nurse not only confirms he followed his profession's highest ethical standards, but also sets an example that all health professionals should follow" (Physicians for Human Rights 2015).

6.11 Addressing Dual Loyalties

We should take note of recommendations from the Defense Health Board. It is a federal advisory committee to the Secretary of Defense offering guidance on health matters. According to Lisa Rapaport,

> A review last year by the board noted that military doctors often face "dual loyalty" conflicts between their ethical responsibilities and their obligations to the military. The board recommended that the DoD ensure its policies, guidelines and instructions let clinicians make the patient their top ethical priority (Rapaport 2016).

This is in line with the WMA Declaration of Malta on hunger strikes. Specifically with regard to balancing dual loyalties:

> Physicians attending hunger strikers can experience a conflict between their loyalty to the employing authority (such as prison management) and their loyalty to patients. In this situation, physicians with dual loyalties are bound by the same ethical principles as other physicians, that is to say that their primary obligation is to the individual patient. They remain independent from their employer in regard to medical decisions (World Medical Association 2017).

Let us now turn to our next example of degrading treatment, that of waterboarding.

6.12 Four: Waterboarding

From all reports being subjected to waterboarding is utterly frightening. We see this in the testimony of some who were subjected to it in a training exercise, as reported by Jessica Schulberg (2018).

- George Wolske, former Navy air crewman, waterboarded in 1969 during Survival, Evasion, Resistance, and Escape (SERE) training:

  > I tell you … it was, perhaps the most searing, burning learning experience I've ever been through. The whole process is introduced and taught as an element of torture. And that it

is .... You are actually brought to the point of thinking you're going to drown. .... They put you down on a board, put you on your back, strap you down, you can't move. They begin pouring water on your face. It's going up your nose, you can't breathe. In that flash of a moment, you recognize that the only thing that really matters in life is oxygen. You can do without a whole lot of stuff — but if you can't breathe, you are going to die.

- Jeromy Shane, former Army interrogation instructor, waterboarded in 2003 during SERE training:

    It was the worst thing I've ever felt .... Your body thinks you're drowning and stops acting appropriately .... It's physically painful. I don't know if you've ever gotten water up your nose while you're swimming. It's that over and over again until who's doing it makes it stop. You've got water in your lungs, your brain is on fire, your nasal cavity is on fire, your throat is completely swollen up.

- Malcolm Nance, retired interrogator, waterboarded in 2006 at a SERE school:

    Waterboarding is slow motion suffocation with enough time to contemplate the inevitability of black out and expiration — usually the person goes into hysterics on the board. For the uninitiated, it is horrifying to watch and if it goes wrong, it can lead straight to terminal hypoxia. When done right it is controlled death .... It is an overwhelming experience that induces horror, triggers a frantic survival instinct. As the event unfolded ... I was being tortured.

- Charlie Thompson, former naval aviator, wrote in 2012 about being waterboarded at a SERE training facility:

    When the restraints were removed, I rolled off the board and onto my hands and knees. I could not have stood up if my life depended on it. And I remained there vomiting water, and gasping for breath when I was asked "You bomb peace loving people in city ... you kill women and children?" I shook my head in the affirmative. At that point I'd have confessed to being the reincarnation of Jack the Ripper if it meant I could avoid another session.

Looking over these four descriptions of being waterboarded, it's safe to conclude that it is a terrifying experience and one not likely to be beneficial in an interrogation. As Charlie Thompson indicated, he'd confess to virtually anything if it would put an end to such brutality. This opinion is shared by Sen. John McCain, who suffered years-long torture in the Vietnam War. McCain said, "Waterboarding is torture, period. I can ensure you that once enough physical pain is inflicted on someone, they will tell that interrogator whatever they think they want to hear" (*Fox News* 2009).

You can see from the above descriptions that waterboarding could be regarded as a near death experience and not just painful to undergo. Although it is often referred to as simulated drowning, it may be more accurate to call it *forced* drowning.

Variations of waterboarding have also be utilized, as, for example, with Mohamed Ben Soud:

Mohamed Ben Soud cannot say for certain when the Americans began using ice water to torment him. The C.I.A. prison in Afghanistan, known as the Salt Pit, was perpetually dark, so the days passed imperceptibly. The United States called the treatment "water dousing," but the term belies the grisly details.

Mr. Ben Soud, in court documents and interviews, described being forced onto a plastic tarp while naked, his hands shackled above his head. Sometimes he was hooded. One C.I.A. official poured buckets of ice water on him as others lifted the tarp's corners, sending water splashing over him and causing a choking or drowning sensation. He said he endured the treatment multiple times (Apuzzo et al. 2016).

Bringing people to the edge of death may elicit credible information —actionable intelligence—but that seems questionable. And when we think about its repeated use on high-profile suspects—e.g., Abu Zubaydah 83 times, Khalid Sheikh Mohammed 183 times—it's not surprising that videotape evidence was destroyed. That suggests an awareness that a line had been crossed—morally and legally. Shadows of culpability.

Furthermore, this was accomplished with the knowledge of now-CIA director, Gina Haspel (Schulberg 2018). Obviously the motivation was to leave as little documentation as possible that would disclose the process and its effect on the victim. So much for thinking of waterboarding as a legitimate tactic in an interrogation. Instead, destroying the evidence is an attempt to avoid scrutiny and could be regarded as an admission of wrongdoing.

That waterboarding is an extreme form of abuse has not prevented its use over the years, as historical records tracing back at least as far as the 1800's indicate. For example, the "shower-bath" was used in 1852 at Auburn State Prison in New York. As reported in *The New York Times* archives, "the fall of water upon the head often produces unconsciousness" (February 1852a).

This water treatment was also used on prisoner Henry Hagan at New York's Sing Sing prison in 1852. In Hagan's case, they first shaved his head and then poured up to 12 barrels of water over it (*The New York Times* April 1852b). That this form of abuse has endured for over 150 years boggles the mind.

Waterboarding (AKA the "water treatment") was applied to al Qaeda suspect Abd al-Rahim al-Nashiri in 2002 at a secret base in Thailand. CIA director Gina Haspel was chief of the base at the time and apparently wrote or authorized the cables detailing the treatment (Barnes and Shane 2018). Her involvement in waterboarding suspects did not prevent the US Congress from confirming her appointment.

Evidently her claim that she would no longer authorize its use was sufficient to put any doubts to rest. That said, "Haspel's critics argued she still wouldn't say if she thought the interrogation program was immoral" (*CNN* 2018). Given comments by the current president endorsing waterboarding, the door may not yet be closed on the practice being put to use in the future. Even with the support of politicians and a percentage of the public opinion, key concerns remain.

Bioethicist-physician Matthew K. Wynia turns the spotlight on doctors:

> The real question at hand is how to hold to account the medical personnel needed to support waterboarding detainees. The method is so dangerous, according to the SERE trainer, that it can be carried out with any degree of safety only if a medical team is at hand. …
> "Presumably" medical teams have monitored the waterboarding of detainees by the CIA … Beyond ethics, participation in waterboarding could also put these physicians in significant legal jeopardy as parties to a war crime (2008, 11).

Let us now turn to our last example of degrading and dehumanizing treatment.

6.13 Five: Humiliation

The use of humiliating practices is a direct attack on the dignity of the victims. It instills a sense of hopelessness, helplessness, and stress with long-reaching consequences. Hooding, forced nudity, force-feeding, and waterboarding either singularly or in combination are cruel and dehumanizing. So too is humiliation. Detainees and prisoners have been subjected to three forms of humiliation that merit concern. They are forced shaving, prolonged use of diapers, and sexual abuse. Let us look at each one.

First consider forced shaving. The very idea of nonconsensual shaving, other than for a medical reason like a lice infestation, calls to mind the practice of shaving off beards in the Holocaust. Since our hair is inherently part of our identity, shaving not only affects our sense of self, it also affects how others see us.

Forced shaving could be considered a hate crime, an act of aggression born of prejudice and disdain. The victim cannot refuse and, therefore, is at the mercy of others; others who wield their power to disorient and strip away the self control of the victim. We see this in the following case:

> A former FBI OSC [Office of Special Council] at [Guantanamo Bay] said that in 2002 the FBI agent interviewing detainee Ghassan Abdullah Al-Sharbi (#682) told him that Al-Sharbi should have his beard shaved. The agent told the OSC that Al-Sharbi's beard was down to his waist and he was getting too much respect on the cell block for this.
>
> The agent recommended that Al-Sharbi be shaved in order to reduce his influence and the level of respect he was receiving from the other detainees on the cell block. ... The OSC gave his consent and shortly thereafter Al-Sharbi's beard was shaved (Human Rights UC Davis A).

Presumably, Al-Sharbi suffered a loss of influence and respect, as was intended. However, such humiliation may not accomplish that goal, given that forced shaving may actually elicit sympathy and indignation. Consequently, it could heighten, not lessen, respect and influence. In any case, to inflict forced shaving without a legitimate cause crosses an ethical boundary that should never have been authorized.

We see this also with the shaving of Abu Zubaydah while in CIA custody. Zubaydah testified that:

> "They kept shaving my head and my face with an electrical razor and they did it in such a quick and violent manner," wrote Zubaydah ... He began vomiting and a nurse was sent in. "I couldn't cover my genitals in an appropriate manner." "Why are you naked?" she asked. "Ask them," he replied. She said: "I'll see what I can do."
>
> Perhaps she complained or was part of the pageant, he could not decide, but the guards returned and clothed him. "Praise God, I am finally able to cover my genitals," he thought. But it did not last long. "Someone started screaming ... and quickly cutting my clothes. I felt at that moment he was cutting my skin." He was also shaved again, "like you shave a sheep and not a human being" (Levy 2017).

His description of being shaved "like you shave a sheep" indicates how humiliating is this practice. How can it be justified? Where is the outrage?

6.14 Prolonged Diapering

A second example of humiliation used on detainees is prolonged diapering. In some cases diapers were not changed for days or weeks at a time, at great discomfort. Diapers obviously have a legitimate purpose with adults, particularly in a medical situation where access to toilet facilities is not an option. However, being forced to wear a diaper for days on end when there is no medical need is demeaning and inhumane.

In addition, over time it becomes a health issue, putting the person at risk of sores and infection. That health professionals should participate in the unnecessary use of diapers is an affront to their fiduciary duties and their relationship to their patients. By no means is prolonged diapering an act of caregiving.

Physicians for Human Rights notes that this practice has been authorized from the top and, therefore, is not the act of a handful of guards. It is a disgusting, unhealthy form of punishment. The International Red Cross reports on detainees in CIA custody:

> Detainees were placed in diapers and denied access to a toilet for prolonged periods of time. For example high-value detainees in CIA custody replace diapers for transport and not allowed to go to the toilet—if necessary they were obliged to urinate or defecate in the diaper.
>
> Some were compelled to wear a diaper for prolonged periods while shackled in a stress standing position. In addition to the helplessness and sense of humiliation and autonomy this would create there's also the physical risk of skin infection, skin breakdown, ulcers and urinary tract infections (Physicians for Human Rights 2009).

It is safe to conclude that prolonged diapering is inhumane. And the fact that medical personnel have enabled it is reprehensible.

6.15 Another Form of Humiliation Is Sexual Abuse

The third type of humiliation detainees have been subjected to is sexual abuse. This could take a number of forms—including nude photography, fondling, rape or sodomy and forcing the victim to commit sexual acts like masturbation or having sex with others. Two examples illustrate the range that sexual abuse can take; first is at Abu Ghraib and, secondly, at Yemeni prisons.

6.16 The Case of Abu Ghraib

Sexual misconduct at Abu Ghraib was widespread. Physicians for Human Rights (PHR) reports on one form this took; namely, the use of women in interrogations:

> These accounts were confirmed to PHR by a source familiar with conditions there. According to the source, in 2003 female interrogators used sexually provocative acts as part of interrogation. For example, female interrogators sat on detainees' laps and fondled themselves or detainees, opened their blouses and pushed their breasts in the faces of detainees,

opened their skirts, kissed detainees and if rejected, accused them of liking men, and forced detainees to look at pornographic pictures or videos. (Human Rights UC Davis B).

That's not all. Female soldiers also smeared what they said was menstrual blood on male detainees. Other forms of sexual abuse at Abu Ghraib, which the extensive photographs shown at Congress and others made public, include forcing nude detainees to wear women's underwear over their heads and faces, posing in sexual positions, simulated or actual sex with other detainees and/or guards, sodomy and rape, including with objects.

Sexual humiliation became a source of entertainment, as seen with forcing nude detainees to form human pyramids in a twisted kind of gymnastics. Women weren't spared either. Female detainees reported on having to bare their breasts and many reported rapes, some resulting in pregnancies.

The reaction of members of Congress to the photographic evidence sums it up: "What we saw is appalling," said Senate Majority Leader Bill Frist (R-Tenn.). "I saw cruel, sadistic torture," said Rep. Jane Harman (D-Calif.) (Babington 2004).

6.17 The Case of Yemeni Prisons

A second example of sexual humiliation reportedly took place in United Arab Emirate (UAE)-run prisons in Yemen in 2017. Specifically,

> Descriptions of the mass abuse offer a window into a world of rampant sexual torture and impunity in UAE-controlled prisons in Yemen …. secret prisons and widespread torture were exposed by an AP [*Associated Press*] investigation [in June 2017] …. The AP has since identified at least five prisons where security forces use sexual torture to brutalize and break inmates.
>
> Witnesses said Yemeni guards working under the direction of Emirati officers have used various methods of sexual torture and humiliation. They raped detainees while other guards filmed the assaults. They electrocuted prisoners' genitals or hung rocks from their testicles. They sexually violated others with wooden and steel poles (Truth Dig 2018).

6.18 Conclusion

Sexual abuse, like the other two types of humiliation we looked at, has left a trail of suffering. That forced shaving, prolonged diapering and sexual humiliation are tactics that were put to use—in some cases widely—indicate an utter disregard for individual dignity. Not only were they tolerated, those with the power to stop them failed to do the right thing.

Psychiatry professor Edmund Howe examines the potential consequences of such practices as these, pointing out the discrepancy between short-term gains and long term harms. He says, "even if harsh methods aren't used, military health professionals' involvement still may be ethically problematic … Depending on what they do, they might use their skills to exploit detainees' vulnerability" (2009, 77).

In line with Kant's second formulation of the Categorical Imperative, the Humanitarian Principle, Howe argues: "They may 'use' detainees too much as primarily means to [their] ends, as opposed to regarding and treating detainees to a sufficient extent as ends in and of themselves" (Howe 2009, 76). Kant would condemn such use. Howe concludes that,

> Even if harsher methods do save more lives in the short-term, they may also result in more lives being lost over the longer run. The use of harsher approaches may, for instance, cause greater animosity and, indeed hatred in others for generations to come. In net effect, this may result in a greater loss of lives" (2009, 77).

There is little doubt that these demeaning practices have brought animosity and resentment in their wake. They have also left the victims with damage that can't be erased—broken men and women. "And then there is the despair of men who say they are no longer themselves. "I am living this kind of depression," said Younous Chekkouri, a Moroccan, who fears going outside because he sees faces in crowds as Guantánamo Bay guards. "I'm not normal anymore" (Apuzzo et al. 2016).

It's not a pretty picture. *The New York Times* reports:

> After enduring agonizing treatment in secret C.I.A. prisons around the world or coercive practices at the military detention camp at Guantánamo Bay, Cuba, dozens of detainees developed persistent mental health problems, according to previously undisclosed medical records, government documents and interviews with former prisoners and military and civilian doctors. Some emerged with the same symptoms as American prisoners of war who were brutalized decades earlier by some of the world's cruelest regimes (Apuzzo et al. 2016).

Is this what we want? Surely not.

Works Cited

Amon, Joseph. 2010. Abusing Patients: Health Providers' Complicity in Torture and Cruel, Inhuman or Degrading Treatment. *Human Rights Watch*. https://www.hrw.org/world-report/2010/country-chapters/global. Retrieved 20 March 2019.

Aoláin, Fionnuala Ní. 2016, June 2. Forced Nudity: What International Law and Practice Tell Us. *Political Settlements Research Program*. http://www.politicalsettlements.org/2016/06/02/forced-nudity/. Retrieved 10 June 2018.

Apuzzo, Matt, Sheri Fink and James Risen. 2016, October 8. How U.S. Torture Left a Legacy of Damaged Minds, *The New York Times*. https://www.nytimes.com/2016/10/09/world/cia-torture-guantanamo-bay.html. Retrieved 10 June 2018.

Babington, Charles. 2004, May 13. Lawmakers Are Stunned New Images of Abuse. *The Washington Post*. https://www.washingtonpost.com/archive/politics/2004/05/13/lawmakers-are-stunned-new-images-of-abuse/1af63486-1724-4485-8f8b-4e857b8fab8a/?utm_term=.fdf5fe69f282. Retrieved 10 June 2018.

Barnes, Julian E. and Scott Shane. 2018, August 10. Cables Detail C.I.A. Waterboarding at Secret Prison Run by Gina Haspel. *The New York Times*. https://www.nytimes.com/2018/08/10/us/politics/waterboarding%2D%2Dhaspel-cia-prison.html. Retrieved 10 November 2018.

Boston University Medical Campus. 2010, February 16. Human Rights & Health Forum: BU Experts to Discuss Hunger Strikes and Force-Feeding of Prisoners Feb. 23. *Boston University*. https://www.bumc.bu.edu/2010/02/16/human-rights-health-forum-bu-experts-to-discuss-hunger-strikes-and-force-feeding-of-prisoners-feb-23/. Retrieved 10 June 2018

Boyd, J. Wesley. 2015, October. Force-Feeding Prisoners Is Wrong. *AMA Journal of Ethics*. https://journalofethics.ama-assn.org/article/force-feeding-prisoners-wrong/2015-10. Retrieved 10 June 2018.

CNN. 2018, May 17. Gina Haspel: Senate Confirms First Female CIA Director. *CNN*. https://www.cnn.com/2018/05/17/politics/gina-haspel-confirmation-vote/index.html. Retrieved 10 June 2018.

Crawford, Monique. 2016, March 29. Shocking Sexual Humiliation Torture at the Hands of the CIA. *Catholic Online, California Network*. https://www.catholic.org/news/national/story.php?id=68161. Retrieved 10 June 2018.

Crosby, Sondra S., Caroline M. Apovian, and Michael A. Grodin. 2007, September. Hunger Strikes, Force-Feeding, and Physicians' Responsibilities. *Journal of the American Medical Association*. https://www.researchgate.net/profile/Caroline_Apovian/publication/6171973_Hunger_Strikes_Force-feeding_and_Physicians%27_Responsibilities/links/0c960518d19265a840000000/Hunger-Strikes-Force-feeding-and-Physicians-Responsibilities.pdf?origin=publication_detail. Retrieved 10 June 2018.

Democratic Underground. 2009, June. Is Obama Condoning Torture by Allowing Force-Feeding at Gitmo? DemocraticUndergound.com. https://www.democraticunderground.com/discuss/duboard.php?az=view_all&address=389x5017717. Retrieved 10 June 2018.

Fay, George. 2004, August 23. Report: Investigation of 205th Military Intelligence Brigade's Activities in Abu Ghraib Detention Facility. *The Torture Data Base, ACLU*. https://www.thetorturedatabase.org/document/fay-report-investigation-205th-military-intelligence-brigades-activites-abu-ghraib. Retrieved 10 June 2018.

Grodin, Michael A. and George J. Annas. 2006, February 9. Hunger Strikers at Guantánamo. *The New York Times*. https://www.nytimes.com/2006/02/15/opinion/hunger-strikers-at-guantanamo-751960.html. Retrieved 10 June 2018.

Harbury, Jennifer K. 2005. *Truth, Torture, and the American Way*. Boston, MA: Beacon Press.

Howe, Edmund. 2009. Further Considerations Regarding Interrogations and Forced Feeding. In *Interrogations, Forced Feedings, and the Role of Health Professionals*, ed. Ryan Goodman and Mindy Jane Roseman. Cambridge: Human Rights Program, Harvard Law School.

Human Rights UC Davis A. K. Forced Shaving. The Guantánamo Testimonials, Center for the Study of Human Rights in the Americas. http://humanrights.ucdavis.edu/projects/the-guantanamo-testimonials-project/testimonies/testimony-of-the-department-of-justice/observations-regarding-specific-techniques/k-forced-shaving. Retrieved 10 June 2018.

Human Rights UC Davis B. Physicians for Human Rights (PHR). Colorado Springs Human Resources Association. http://humanrights.ucdavis.edu/projects/the-guantanamo-testimonials-project/testimonies/other-testimonies/physicians-for-human-rights-phr. Retrieved 10 June 2018.

Iacopino, Vincent. 2015, May. Navy Nurse Press Call, Physicians for Human Rights. http://physiciansforhumanrights.org/library/statements/navy-nurse-press-call.html. Retrieved 10 June 2018.

Irish Times. 2016, January 6. The Torture Centre: Northern Ireland's 'Hooded Men'. *Irish Times*. https://www.google.com/amp/s/www.irishtimes.com/news/crime-and-law/the-torture-centre-northern-ireland-s-hooded-men-1.2296152%3fmode=amp. Retrieved 10 June 2018.

Jackson, Richard. 2007, July. Language, Policy and the Construction of a Torture Culture in the War on Terrorism. *Review of International Studies* 33 (3): 353–371. https://www.cambridge.org/core/journals/review-of-international-studies/article/language-policy-and-the-construction-of-a-torture-culture-in-the-war-on-terrorism/5006D871A4E5550BE5EBA0F10CCEDACF. Accessed 10 June 2018. Retrieved 10 June 2018.

Khamsi, Roxanne. 2006, March 10. Doctors Decry Force-Feeding at Guantanamo Bay. *New Scientist*. http://www.newscientist.com/channel/opinion/dn8828-doctors-decry-forcefeeding-at-guantanamo-bay.html. Retrieved 10 June 2018.

Legislation on Line. 1992. Declaration on Hunger Strikers (Declaration of Malta) (1991, revised 1992). https://www.legislationline.org/documents/id/8591. Retrieved 10 June 2018.

Levy, Adrian. 2017, May 18. The Horrifying True Story of a CIA Waterboarding Victim. *Vice*. https://www.vice.com/en_us/article/4xkp8q/the-horrifying-true-story-of-a-cia-waterboarding-victim. Retrieved 10 June 2018.

Lindh, Frank. 2014, August 28. How John Walker Lindh Became 'Detainee 001'. *The Nation*. https://www.thenation.com/article/how-john-walker-lindh-became-detainee-001/. Retrieved 10 June 2018.

Mazzetti, Mark and Scott Shane. 2008, March 13. Pentagon Cites Tapes Showing Interrogations. *The New York Times*.

McCain, John. 2009, April 20. Interview. *Fox News*. http://www.city-data.com/forum/politics-other-controversies/626405-mccain-reacts-ksm-being-waterboarded-183-a.html. Retrieved 10 June 2018.

NPR. 2004, February 4. Alleged Methods of Ill Treatment, An Excerpt from the [international] Red Cross Report on Prisoner Abuse in Iraq. *NPR*. https://www.npr.org/iraq/redcross/abuse_report.html. Retrieved 10 June 2018.

Physicians for Human Rights. 2009, August. Aiding Torture: Health Professionals' Ethics and Human Rights Violations Revealed in the May 2004 CIA Inspector General's Report.

———. 2015, July 22. Guantánamo Navy Nurse Who Refused to Force-Feed Detainees to Receive Ethics Award. http://physiciansforhumanrights.org/press/press-releases/guantanamo-navy-nurse-who-refused-to-force-feed-force-feed-detainees-to-receive-ethics-award.html. Retrieved 10 June 2018.

Rapaport, Lisa. 2016, January 5. Ethicists Say Military Doctors Shouldn't Force-Feed Prisoners. *Reuters*. https://www.reuters.com/article/us-health-military-ethics/ethicists-say-military-doctors-shouldnt-force-feed-prisoners-idUSKBN0UJ1YZ20160105. Retrieved 10 June 2018.

Rosenberg, Caroline. 2019, April 6. Guantánamo Trials Grapple with How Much Evidence to Allow About Torture. *The New York Times*. https://www.nytimes.com/2019/04/05/us/politics/guantanamo-trials-torture.html?smid=nytcore-ios-share. Retrieved 6 April 2019.

Schulberg, Jessica. 2018, May 9. Here's What Waterboarding Is Really Like, According to People Who Suffered Through It. *Huffington Post*. https://www.huffingtonpost.com/entry/what-waterboarding-is-really-like_us_5ab3b4bae4b008c9e5f4d6b5. Retrieved 10 June 2018.

The Economist. 2012, September 22. A Journey Into Hell. https://www.economist.com/the-americas/2012/09/22/a-journey-into-hell. Retrieved 10 June 2018.

The Guardian (UK). 2006, March 10. Doctors Demand End to Guantanamo Force-Feeding. *Guardian (UK)*. https://www.theguardian.com/world/2006/mar/10/guantanamo.usa. Retrieved 10 June 2018.

The Guardian (UK). 2016, March 28. CIA Photographed Detainees Naked Before Sending Them to Be Tortured. ———. https://www.theguardian.com/us-news/2016/mar/28/cia-photographed-naked-detainees. Retrieved 10 June 2018.

The New York Times. 1852a, February 18. A Howard Wanted. *The New York Times* archives. https://timesmachine.nytimes.com/timesmachine/1852/02/18/74856547.pdf. Retrieved 10 June 2018.

———. 1852b, April 6. Extra-Judicial Punishment. *The New York Times* archives. https://timesmachine.nytimes.com/timesmachine/1852/04/06/74858474.pdf. Retrieved 10 June 2018.

Truth Dig. 2018, June 20. Drawings Expose Sexual Torture in UAE-Run Prisons in Yemen. *Truth Dig*. https://www.truthdig.com/articles/drawings-expose-sexual-abuses-in-uae-run-prisons-in-yemen/. Retrieved 12 July 2018.

US Department of Justice. 2012, January 6. An Updated Definition of Rape. *Justice.org*. https://www.justice.gov/archives/opa/blog/updated-definition-rape. Retrieved 6 April 2019.

World Medical Association. 2017, October 15. WMA Declaration of Malta on Hunger Strikers *WMA*. https://www.wma.net/policies-post/wma-declaration-of-malta-on-hunger-strikers/. Retrieved 10 June 2018.

Works Cited

Wynia, Matthew K. 2008, January–February. Policy & Politics: Laying the Groundwork for a Defense against Participation in Torture? *The Hastings Center Report* 38 (1): 11–13. http://www.jstor.org/stable/25165285. Retrieved 10 June 2018.

Xenakis, S.N. 2017, September. Ethics Dilemmas in Managing Hunger Strikes. *The Journal of the American Academy of Psychiatry and the Law* 45 (3): 311–315. https://www.ncbi.nlm.nih.gov/pubmed/28939728. Retrieved 10 June 2018.

Zernike, Kate and David Rohde, 2004, June 8. The Reach of War: Sexual Humiliation; Forced Nudity of Iraqi Prisoners Is Seen as a Pervasive Pattern, Not Isolated Incidents. *The New York Times.* https://www.nytimes.com/2004/06/08/world/reach-war-sexual-humiliation-forced-nudity-iraqi-prisoners-seen-pervasive.html. Retrieved 10 June 2018.

Chapter 7
Solitary Confinement

> *When forces are beyond your control, there's not a lot you can do.*
> —Albert Woodfox.
> *It's an awful thing solitary... It crushes your spirit and weakens your resistance more effectively than any other form of mistreatment.*
> —Sen. John McCain.
> *There is torture of mind as well as body; the will is as much affected by fear as by force. And there comes a point where this court should not be ignorant as judges of what we know as men.*
> —Justice Felix Frankfurter.

Overview

In Part II we look at the boundaries of torture and the ethical issues they bring to light. In this chapter we turn to what many consider the worst form of torture, solitary confinement. A disaster in terms of a method of rehabilitation (as intended by the Quakers who put it to use years ago), solitary confinement has brought about any number of harms and no apparent benefits.

The days of the Quakers thinking solitary confinement purifies the soul are long gone. In its place is we find a practice that causes both mental and physical suffering with long-term damage. We will look at five cases to get a sense of the potential harm of solitary. They include a lawyer, a detainee, two hostages, three prisoners, and three men on death row. The picture they present is truly alarming.

Each of these cases indicate why it should be banned or restricted to a matter of hours if absolutely necessary. We then look at the examples of countries that have made changes and, within the U.S., the actions of Rick Raemisch, the Colorado governor who put an end to long term solitary in his state.

7.1 Introduction

It is no mistake on Nel Noddings' part to consider isolation a manifestation of evil. The utter cruelty of confining another person to a cell, cage, or box boggles the mind. That it can go on for years or even decades is unimaginable. Where was the outcry when Zacarias Moussaoui was sentenced to life in prison in solitary confinement with no possibility of parole for his role in the attacks on 9/11?

The fantasy of the Quakers who thought solitary confinement would purify the soul and be an effective form of rehabilitation has been shown to be groundless. That it has continued to the present day and is commonly used in prisons and detention centers indicates how misguided is this policy and how urgent is the need to put an end to it.

Let's assume for the moment that there could be therapeutic benefits to a short period of isolation. If that were the case, we would need to clarify what they are and whether they outweigh the long term risks. The practice needs to be assessed by medical professionals to fully grasp the potential harms and consider the alternatives.

Cyrus Ahalt and Brie Williams (2016) are right on track in raising this concern: "Aside from conducting some important studies that have linked even relatively brief isolation to worsening mental health, the medical community has been largely absent from the national debate over solitary confinement. That absence is conspicuous."

It's conspicuous and it needs to change. This is particularly so, given the rampant misuse of the practice. We see this with the number of minor infractions that result in "segregation" (= solitary).

7.2 The Misuse of Solitary

Joshua Manson cites examples from a United Nations' report.

> In both the U.S. immigration and detention and federal prison systems, for example, people can be held in solitary confinement for the sole reason that they will be released, removed, or transferred within 24 hours. Also under U.S. federal law, people can be held in solitary confinement if they are HIV-positive and there is "reliable evidence" that they may "pose a health risk" to others.
>
> In California, people may be subject to isolation just because they are "a relative or associate of a staff member." In Pennsylvania, people may be subject to solitary confinement, against their will and consent, because they are "at a high risk of sexual victimization" if there exists no "alternative means of separation from the likely abuser." Also in Pennsylvania, people may be subject to solitary confinement simply because "there is no other appropriate bed space" (Manson 2016).

And so on. You don't have to be a danger to others to end up in solitary confinement, contrary to what we might think. This is particularly disturbing in light of the potential harms of being placed in solitary for even short periods. It can cause long-lasting trauma, mental suffering, and psychological harm.

Solitary can also heighten the risk of suicide. We see this, for example, with the case of Jack Letts. A British Canadian man who went off to Syria to join ISIS, Letts was captured and held in prison by Kurdish authorities. He ended up in solitary.

> He told the [Canadian] consular official that he had attempted suicide after his first month in solitary confinement but was found in time by his Kurdish guards. "I started to go insane and talk to myself and I thought dying was better than my mother seeing me insane," Letts said. "So I tried to hang myself." He has since been allowed to live in a cell with other prisoners (Brewster 2018).

Jack Letts's reaction to solitary matches the experience of one after another after another. We can see this in the following five cases—the use of solitary on a lawyer, a detainee, two hostages, three prisoners, and three on Death Row. Each case paints a bleak picture of the use of solitary confinement.

7.3 First Case: Lawyer Xie Yanyi

Xie Yanyi had spent much of his career in China representing clients in sensitive legal cases involving victims of official corruption, police violence or religious persecution. John Sudworth (2017) of the BBC reports on China's "war on law":

> The crackdown on China's already beleaguered human rights field began in mid 2015, halfway through Xi Jinping's first term. Now, anointed in office for a second term by the Communist Party Congress that ended this week, it stands as one of his most gloomy legacies. In total, more than 300 lawyers, legal assistants and activists have been brought in for questioning, with more than two dozen pursued as formal investigations.

Xie Yanyi, one of those arrested, said he didn't see sunlight for 6 months and that he was kept in a stress position and crouched on a low stool from 6:00 in the morning until 10:00 at night. He elaborates:

> After 15 days like this, [he said], his legs went numb and he had difficulty urinating. At times he was denied food and was beaten … But harder to bear was the time spent in solitary confinement, he asserts. "I was kept alone in a small room and saw no daylight for half a year. I had nothing to read, nothing to do but to sit on that low stool." In his view, "People could go mad in that situation. I was isolated from the world. This is torture—the isolation is more painful than being beaten" (Sudworth 2017).

7.4 Second Case: Detainee Jose Padilla

Detainee and American citizen Jose Padilla was held over 3 years in solitary confinement. The conditions were harsh: His windows were blacked out; there was no clock or calendar and only a steel platform to sleep on (no mattress) before having access to legal counsel (Sontag 2006).

Two psychiatrists and a psychologist who conducted examinations of Padilla contend that both his extended detention in solitary and interrogation left him with severe mental disabilities. All three say he may never recover. In their opinion, "Padilla's ordeal in the brig was so psychologically unsettling that it has left him terrorized. Any reminder of the ordeal through questions by his lawyers or others triggers a recurrence of the disorganizing terror Padilla experienced" (Richey 2007).

7.5 Third Case: Hostages Terry Anderson and Frank Reed

Journalist Terry Anderson was covering Lebanon's civil war of 1975–1990 when he was kidnapped along with other foreigners. He was released in 1991 after being held prisoner in an underground dungeon for six-and-a-half years (History.com 2009).

Fellow hostage Frank Reed, an American private-school director, was held in solitary confinement for 4 months before being put in with Anderson. Surgeon and public health commentator Atul Gawande reports on Reed's condition: "By then, Reed had become severely withdrawn. He lay motionless for hours facing a wall, semi-catatonic. He could not follow the guards' simplest instructions. Released after three and a half years, Reed ultimately required admission to a psychiatric hospital" (2016).

Anderson didn't fare so well either:

> After a few months without regular social contact … he started to lose his mind. He talked to himself. He paced back and forth compulsively, shuffling along the same six-foot path for hours on end. Soon, he was having panic attacks, screaming for help. He hallucinated that the colors on the walls were changing. He became enraged by routine noises—the sound of doors opening as the guards made their hourly checks, the sounds of inmates in nearby cells. After a year or so, he was hearing voices on the television talking directly to him. He put the television under his bed, and rarely took it out again (Gawande 2009).

7.6 Fourth Case: Prisoners Robert King, Herman Wallace, and Albert Woodfox, AKA "The Angola Three"

The three Black Panther-activists were convicted in the early 1970's of a prison guard's murder (they claimed they were framed). Each spent decades in solitary—King 29 years, Wallace 41 years, and Woodfox 43 years, the longest in US history. King commented on the conditions:

> The cells were pretty bare, and they were maybe about … 3 feet wide and about 6 feet long. It was almost like it was in a tomb, and there was a slab of concrete that you slept on. You ate three meals a day — you had two slices of bread each meal. During the wintertime, you froze, and during summertime, you were overheated. But in any event, you were starved (Kennedy 2016).

Wallace died 3 days after his release in 2013. In a February 2016 interview with *The Guardian (UK),* Woodfox remarked on the brutality of solitary:

Some of the guys found the pressure so great that they just laid down in a fetal position and stopped communicating with anybody. I've seen other guys who just want to talk and make noise, guys who want to scream. Breaking up manifests itself in any number of ways in individuals" (Pilkington 2016).

As for himself, Woodfox observed that,

The panic attacks started with sweating. You sweat and you can't stop. You become soaking wet—you are asleep in your bunk and everything is soaking wet. Then when the claustrophobia starts it feels like the atmosphere is pressing down on you. That was hard. I used to talk to myself to convince myself I was strong enough to survive, just to hold on to my sanity until the feeling went away (Pilkington 2016).

7.7 Fifth Case: Death Row Convicts Marcus Hamilton, Winthrop Eaton and Michael Perry

Convicted of first-degree murder in Louisiana and sentenced to death, all three men have been in solitary confinement for over 25 years. They aren't the only ones: According to Liam Stack (2017), 71 prisoners sentenced to death in Louisiana are being held in solitary confinement at the Angola state penitentiary.

They spend 23 hours a day in windowless concrete cells that measure 8 by 10 feet. They are allowed to leave the cell for one hour each day to shower, make phone calls or walk along the tiered walkway beside their cells. Three times a week, they can use that hour to go outside to sit in a small outdoor cage (Stack (2017).

Evidently "a lot, if not most" of the states that have the death penalty house their death row inmates in solitary confinement according to Betsy Ginsberg, the director of the civil rights clinic at the Benjamin N. Cardozo School of Law (Stack 2017).

The damaging effects of prolonged solitary confinement on the mental and physical health of prisoners have been well established, as Stack points out. Social psychologist Craig Haney notes that even those with no history of mental illness were engaged in a "constant, ongoing struggle to maintain their sanity." In addition, "These conditions predictably can impair the psychological functioning of the prisoners who are subjected to them," he [observed]. "For some prisoners, these impairments can be permanent and life-threatening" (Stack 2017).

7.8 Overview of Issues

On any given day, approximately 80,000 prisoners are being held in solitary confinement in the United States alone. That's not all. According to the Bureau of Justice Statistics, over the course of a year, nearly 1 in 5 U.S. prisoners (= 20%) spent time in solitary confinement. This adds up to 400,000 people who are incarcerated are also in solitary during a given year (Ahalt and Williams 2016). With such staggering numbers, a lot more needs to be done to justify this practice.

Others convey similar sentiments and raise important concerns. For instance, it's not only prisoners who are subjected to solitary confinement. Immigrants in detention facilities are also at risk.

John V. Kelly, US Department of Homeland Security Acting Inspector General points out that, "Staff did not always treat detainees respectfully and professionally, and some facilities may have misused segregation" (Sacchetti 2017). He cites the example of a detainee reported being locked down for multiple days for sharing coffee with another detainee (Office of Inspector General, Homeland Security 2017). This is clearly an overreaction to a minor offense.

7.9 The Harms of Solitary

Misuse is one concern. So are the extreme conditions and the resulting psychological effects. Social scientist Craig Haney undertook a two-year study of inmates who were subjected to prolonged solitary confinement at Pelican Bay prison in California. He states that,

> Sealed for years in a hermetic environment — one inmate likened the prison's solitary confinement unit to "a weapons lab or a place for human experiments" — prisoners recounted struggling daily to maintain their sanity.
>
> They spoke of longing to catch sight of a tree or a bird. Many responded to their isolation by shutting down their emotions and withdrawing even further, shunning even the meager human conversation and company they were afforded (Goode 2015).

Haney, said that, after months or years of complete isolation, many prisoners begin to lose the ability to initiate behavior of any kind. In extreme cases, "prisoners may literally stop behaving," thus becoming "essentially catatonic" (Gawande 2009).

In addition, almost 90% of these prisoners had difficulties with "irrational anger,"as Terry Anderson notes about his own experience. In contrast, only 3% of the general population are said to suffer extreme anger. Not surprising, Haney claims that many prisoners in solitary become consumed with revenge fantasies.

This is a far cry from the rehabilitation that Quakers sought in turning to solitary confinement. Rather than a time to cool off, solitary seems more likely to bring emotions to the boiling point. Just look at some of the harms that have been documented.

Atul Gawande (2009) reports that studies show diffuse slowing of brain waves after a week or more of solitary confinement. He cites the 1992 case of 57 prisoners of war whose EEGs showed brain abnormalities months after a six-month detention in Yugoslavia. Such evidence indicates solitary confinement has both physical as well as psychological harms.

7.10 Abusive and Aggressive Forms of Solitary

If things aren't bad enough when people are confined to 8 × 10′ cells, there have been even more alarming cases. A grotesque turn of events has been the use of boxes (for example with Abu Zubaydah) to cram suspected terrorists into. Yes, boxes. Presumably this is to encourage them to be more forthcoming; as if subjecting them to such miserable conditions would inspire them to divulge useful information. The result is that their confinement is more physically taxing than with the use of rooms or prison cells.

Such an extreme has been accepted if not sanctioned at the highest level. In fact, psychologist-masterminds Drs. Mitchell and Jessen recommended using such "aggressive techniques" as part of the interrogation process on so-called high-value captives. This included detainees Abu Zubaydah and Mohamed Ben Soud being placed in wooden boxes with holes poked in them to allow air flow (Fink and Risen 2017). They were then forced into crouched positions, unable to stand up or move about. The Senate Intelligence Committee Report details Zubaydah's confinement to a box:

> He spent a total of 266 hours (11 days, two hours) in the large (coffin size) confinement box and 29 hours in a small confinement box, which had a width of 21 inches, a depth of 2.5 feet, and a height of 2.5 feet... According to the daily cables ... Abu frequently "cried," "begged," "pleaded," and "whimpered," but continued to deny that he had any additional information (2014, 52–53).

To add to the pressure, then US District Attorney Jay Bybee authorized via a memo the use of insects placed in the box to exploit the victim's phobias and fears. The memo permitted Zubaydah to be kept in a dark, confined space small enough to restrict his movement (*CNN* 2009).

> In addition, putting a [so-called] harmless insect into the box with Zubaydah, who "appears to have a fear of insects," and telling him it is a stinging insect would be allowed, as long as Zubaydah was informed the insect's sting would not be fatal or cause severe pain. "If, however, you were to place the insect in the box without informing him that you are doing so ... you should not affirmatively lead him to believe that any insect is present which has a sting that could produce severe pain or suffering or even cause his death," the memo said (*CNN* 2009).

7.11 Descriptions and Denials

The five cases of solitary confinement are disturbing enough when looking at the individuals affected, not to mention those like Abu Zubaydah and Ben Soud, who were crammed into boxes. It is also important to consider the scope of the use of solitary and how it can be obscured or hidden by the way it is described. We see this in the following mixture of linguistic manipulation and denial:

> "The Bureau [of Prisons] does not recognize the term solitary confinement. Therefore, the Bureau does not have a definition or reference to provide," the BOP told investigators with the DOJ [Department of Justice] Office of the Inspector General.
> One former BOP official told investigators that "solitary confinement does not exist" within the federal prisons system. But the inspector general's office said the Bureau of Prisons is just arguing semantics (Reilly 2017).

Moreover, as Law Professor Joseph Margulies points out, there is no upper bound on holding a detainee in solitary confinement or being subjected to interrogations (2006, 107). To get a sense of the enormity of the problem in the US, look at the numbers reported in July, 2017 by the US Department of Justice on the Federal Bureau of Prisons:

> As of June 2016, of the 148,227 sentenced inmates in the BOP's [Federal Bureau of Prisons'] 122 institutions, 9,749 inmates (7 percent) were housed in its three largest forms of Restricted Housing (RHU) ... Although the BOP states that it does not practice solitary confinement, or even recognize the term, we found inmates, including those with mental illness, who were housed in single-cell confinement for long periods of time, isolated from other inmates and with limited human contact. ...
> Although the BOP generally imposes a minimum amount of time that inmates must spend in RHUs, it does not limit the maximum amount of time ... As a result, inmates, including those with mental illness, may spend years and even decades in RHUs (Office of the Inspector General, DOJ 2017).

7.12 The Use of Euphemisms

One of the ways that solitary confinement was made more palatable was through distorting language. Whitewashing it with euphemisms like "disciplinary segregation," "administrative segregation" or "Segregated Housing Units (SHU)" does not alter the fact that such isolation is harmful and its extensive use alarming. With such labeling, they can avoid admitting that they practice solitary confinement. To what advantage? Colin Dayan asserts, "Since prison officials claim that these units are non-punitive, they are difficult to fight" (2007, 54).

The Washington Post Editorial Board (2017) responded to the Department of Justice report by declaring solitary confinement to be torture. "Restrictive housing placement is the bureau's preferred terminology," they stated, "but the report released last week made clear that is a semantic dodge." *The Post* pointed out that:

> Among those forced to languish alone were inmates with serious mental illness, some of whom were isolated for more than five years. A number of state prison systems have taken steps to limit or end their use of solitary confinement because of mounting evidence of its detrimental effects.
> The Inspector General cited research that isolation can cause anxiety, depression, anger, paranoia and psychosis among prisoners. "You have no contact, you don't speak to anybody, and it's a form of torture on some level," a psychologist at one prison told investigators.

7.13 Where Is the Outrage?

Given this is the case, it is puzzling how little outcry there is among medical personnel and the general public. The solution? *The Post* calls it:

> Ending such barbarity is not only morally correct but also has practical benefits in improving public safety. Prisoners subjected to solitary confinement have difficulty reentering society and are more likely to re-offend (2017).

Colin Dayan observes that solitary confinement was once considered the most severe deprivation and now its use is normalized (2007, 54). However, change is finally afoot. We see this in the practice of solitary confinement being banned in the state of Colorado. Rick Raemisch, the Executive Director of the Colorado Department of Corrections, explains:

> Long-term solitary was supposed to be rehabilitative, but it did not have that effect. Studies have found that inmates who have spent time in solitary confinement are more likely to reoffend than those who have not. Data shows that prisoners in solitary account for about half of all prison suicides; self-harm is also more common in solitary units than in less-restrictive ones.
>
> One social psychologist even found that the degree of loneliness experienced by people in solitary is matched only by that of terminal cancer patients. In addition, solitary confinement was intended to be a last resort for those who were too violent to be in a prison's general population. But then we gradually included inmates who disrupted the efficient running of an institution. In other words, inmates could be placed in solitary for almost any reason, and they were (Raemisch 2017).

7.14 Working for Change

Raemisch came to see that major changes were in order. He noted how he benefited from listening to the perspectives of others. This came about in 2015 when he assisted the State Department with other United Nations countries in modernizing international standards for the treatment of prisoners—known as the Nelson Mandela Rules. It was decided, Raemisch (2017) reports, that keeping someone for more than 15 days in solitary was torture.

This conclusion was echoed in a report of the Human Rights Clinic at the University of Texas School of Law. "Its authors concluded," reports Jacey Fortin, "that solitary confinement in Texas violates international human rights standards and amounts to a form of torture" (2017). The bottom line is that:

> We agree that if somebody commits a crime, they should receive a punishment… But we say, at the same time, that once you impose a punishment on a person, you need to keep treating that person with humanity and with dignity (Fortin 2017).

Raemisch would concur. As he puts it, "There now is enough data to convince me that long-term isolation manufactures and aggravates mental illness. It has not solved any problems; at best it has maintained them. That's why, in September, Colorado ended the practice" (2017).

This decision is significant; others should follow their example. We need to recognize that its widespread use in the U. S. doesn't make it right. Many other countries are in the same boat in the use and misuse of solitary.

> In an October17, 2016 UN Special Rapporteur on Torture, Juan E. Méndez, presented a report to the General Assembly detailing and comparing solitary confinement practices around the world. He declared that solitary confinement may amount to cruel, inhuman, or degrading treatment and in some cases torture, and may thus, under certain conditions, be prohibited under international law.
>
> In that 2011 report, Méndez further called for a categorical ban on subjecting juveniles and people with mental illness to solitary confinement, and to end the practice of prolonged and indefinite solitary confinement (Solitary Watch 2016).

In contrast, Norway is at the other end of the spectrum in its use of solitary confinement. As Ahalt and Williams (2016) indicate, Norwegian leaders realized that the old model simply did not work, that it did more harm than good and the time had come for its replacement. Norway's approach stands as a dramatic contrast with the systemic overuse of solitary.

7.15 What Role Can Doctors Play?

Doctors have a role to play in this. The Norwegian model is worth serious consideration. Ahalt and Williams (2016) offer their recommendation—one it behooves health professionals to take. "We believe the health professions have a responsibility to work with criminal justice policymakers to assess the risk of health-related harm underlying correctional practices such as solitary confinement. Fortunately, a compelling model for such a partnership exists." This they saw practiced at a Norwegian Maximum-security prison. It housed 250 men, many serving long sentences for violent crimes.

> We asked the warden how many prisoners were being held in isolation and were surprised to hear his answer: one. That morning, a prisoner had trashed his cell. Guards, using motivational interviewing strategies, tried unsuccessfully to defuse the situation. The man was now cooling off in isolation under the close supervision of a health care team. "As soon as he calms down," the warden told us, "he will return to the general population." We asked how long that might take. "Tonight?" the warden guessed. "Tomorrow, certainly" (Ahalt and Williams 2016).

In comparison:

> In a U.S. prison of the same size, we would expect to find 25 prisoners in solitary, and roughly half of them would be confined for more than a month. But here there was only one serving less than 1 day, with enhanced attention rather than minimal human contact.

The Norwegian criminal justice leaders we spoke with told us that 20 years ago their correctional system resembled ours: overcrowded and violent, with frequent use of solitary confinement. Then, motivated by prison riots, Norwegian leaders undertook broad reform (Ahalt and Williams 2016).

Norway is not alone casting a critical eye on solitary in trying to make changes with this outdated and harmful model. What is going on in Britain also deserves our attention. Jean Casella (2015) of Solitary Watch reports on controls that have been put in place:

> Britain has a long history of using solitary confinement. In 1842, it mimicked the United States in opening a prison, ... devoted to isolating prisoners so they could contemplate their sins. The practice was largely abandoned on both sides of the Atlantic once it became clear that solitary led to madness, not penitence. Prison isolation was revived in the late 20th Century, but never on the scale seen in the United States ....
>
> Today, a series of rules allow the short-term use of solitary confinement, but are meant to check its overuse and abuse. Prison governors (wardens) may place individuals in what are officially called "care and separation" units to preserve the "good order and discipline" of the prison. But after 72 hours, their continued segregation requires the approval of not only the governor, but—as confirmed in a recent court decision—the Cabinet Secretary for Justice. With the necessary approvals, terms in solitary can stretch longer, but must be reviewed and renewed every 14 days.

With the additional oversight and a high-level approvals process, the abuse of solitary confinement should be curtailed. Moreover, this model allows for more transparency and thus has less likelihood of misuse.

7.16 Conclusion

There is little doubt that solitary confinement is cruel and unusual punishment and, for any extended period, is a form of torture. It has virtually no value as an instrument of rehabilitation. That it is not been banned or subject to strict limitations in terms of use and duration is simply deplorable.

Both Norway and Britain have moved to a different model of addressing behavioral or other issues that made solitary seem a legitimate option. We need to give close attention to the approaches they have taken and consider the range of alternatives to solitary.

Although it seems unreasonable to have oversight by a governor, a director of justice or the like, an oversight committee could be employed to help end the abuses taking place with solitary confinement.

One option might be to use a review board such as a Bioethics committee or a variation drawing upon medical professionals, clergy, and members of the community. In other words, we need more transparency about what is taking place and more controls on what is being done.

Works Cited

Ahalt, Cyrus, and Brie Williams. 2016, May 5. Reforming Solitary-Confinement Policy — Heeding a Presidential Call to Action. *New England Journal of Medicine* 374: 1704–1706. http://www.stopsolitaryforkids.org/wp-content/uploads/2016/05/NEJM-Solitary-May-2016-p1601399.pdf. Retrieved 5 June 2018.

Brewster, Murray. 2018, February 8. Alleged ISIS Operative 'Jihadi Jack' Begs Canada to Let Him Come Here. *CBC News*. http://www.cbc.ca/news/politics/jihadi-jack-isis-consular-1.4526882. Retrieved 5 June 2018.

Casella, Jean, 2015, October 21, "Off the Block," *Solitary Watch* http://solitarywatch.com/2015/10/21/off-the-block/. Retrieved 5 June 2018.

CNN. 2009, April 17. 2002 Memo: Had to Be Intent to Inflict 'Severe Pain' to Be Torture. *CNN*. http://www.cnn.com/2009/POLITICS/04/16/us.torture.documents/. Retrieved 5 June 2018.

Dayan, Colin. 2007. *The Story of Cruel and Unusual*. Boston: MIT Press.

Fink, Sheri and James Risen. 2017, June 20. Psychologists Open a Window on Brutal C.I.A. Interrogations. *The New York Times*. https://www.nytimes.com/interactive/2017/06/20/us/cia-torture.html. Retrieved 5 June 2018.

Fortin, Jacey. 2017, April 26. Report Compares Texas' Solitary Confinement Policies to Torture. *The New York Times*. https://www.nytimes.com/2017/04/26/us/texas-death-row-torture-report.html. Retrieved 5 June 2018.

Gawande, Atul. 2009, March 30. Hellhole. *The New Yorker*. https://www.newyorker.com/magazine/2009/03/30/hellhole. Retrieved 5 June 2018.

Goode, Erica. 2015, August 3. Solitary Confinement: Punished for Life. *The New Yorker*. https://www.nytimes.com/2015/08/04/health/solitary-confinement-mental-illness.html. Retrieved 5 June 2018.

History.com. 2009, December 4. 1991: Hostage Terry Anderson Freed in Lebanon. https://www.history.com/this-day-in-history/hostage-terry-anderson-freed-in-lebanon. Retrieved 5 June 2018.

Kennedy, Merrit, 2016, February 19. Last of 'Angola 3' Released After More Than 40 Years in Solitary Confinement. *NPR*. https://www.npr.org/sections/thetwo-way/2016/02/19/467406096/last-of-angola-3-released-after-more-than-40-years-in-solitary-confinement. Retrieved 5 June 2018.

Manson, Joshua. 2016, October 28. UN Report Compare Solitary Practices in the US and Around the World. *Solitary Watch*. http://solitarywatch.com/2016/10/28/un-report-compares-solitary-confinement-practices-around-the-world/. Retrieved 5 June 2018.

Margulies, Joseph. 2006. *Guantánamo and the Abuse of Presidential Power*. New York: Simon & Schuster.

Office of Inspector General. 2017, December 11. Concerns about ICE Detainee Treatment and Care at Detention Facilities. *Homeland Security*. https://www.oig.dhs.gov/sites/default/files/assets/2017-12/OIG-18-32-Dec17.pdf. Retrieved 5 June 2018.

Office of the Inspector General. 2017, July. Review of the Federal Bureau of Prisons' Use of Restrictive Housing for Inmates with Mental Illness. *US Department of Justice*. https://oig.justice.gov/reports/2017/e1705.pdf#page=1. Retrieved 5 June 2018.

Pilkington, Ed. 2016, February 20. Albert Woodfox Speaks After 43 Years in Solitary Confinement: 'I Would Not Let Them Drive Me Insane'. *The Guardian (UK)*. https://www.theguardian.com/us-news/2016/feb/20/albert-woodfox-angola-3-first-interview-trump-confinement. Retrieved 5 June 2018.

Raemisch, Rick. 2017, October 12. Why We Ended Long-Term Solitary Confinement in Colorado. *The New York Times*. https://www.nytimes.com/2017/10/12/opinion/solitary-confinement-colorado-prison.html. Retrieved 5 June 2018.

Reilly, Ryan J. 2017, July 12. Federal Prisons Officials Claim Inmates Aren't Held in Solitary. DOJ Watchdog Says They Are. *Huffington Post*. https://www.huffingtonpost.com/entry/federal-prison-solitary-confinement-mental-illness_us_59664623e4b005b0fdca5f85?nf5. Retrieved 5 June 2018.

Richey, Warren. 2007, August 13. US Terror Interrogation Went Too Far, Experts Say. *Christian Science Monitor*. https://www.csmonitor.com/2007/0813/p01s03-usju.html. Retrieved 5 June 2018.

Sacchetti, Maria. 2017, December 17. Watchdog Report Finds Moldy Food, Mistreatment In Immigrant Detention Centers. *The Washington Post.* https://www.washingtonpost.com/local/immigration/watchdog-report-finds-moldy-food-mistreatment-in-immigrant-detention-centers/2017/12/15/c97b380a-e10d-11e7-89e8-edec16379010_story.html?utm_term=.1bc140373121. Retrieved 5 June 2018.

Solitary Watch. 2016, October 28. UN Report Compares Solitary Confinement Practices in the U.S. and Around the World. http://solitarywatch.com/2016/10/28/un-report-compares-solitary-confinement-practices-around-the-world/. Retrieved 5 June 2018.

Sontag, Deborah. 2006, December 4. Video Is a Window Into a Terror Suspect's Isolation. *The New York Times*. http://www.nytimes.com/2006/12/04/us/04detain.html. Retrieved 5 June 2018.

Stack, Liam. 2017, March 30. 3 Men on Death Row in Louisiana Sue Over Solitary Confinement. *The New York Times*. https://www.nytimes.com/2017/03/30/us/3-men-on-death-row-in-louisiana-sue-over-solitary-confinement.htm. Retrieved 5 June 2018.

Sudworth, John. 2017, October 25. *China lawyer recounts torture under Xi's 'war on law' BBC news*. Beijing: BBC. http://www.bbc.com/news/blogs-chinablog-41661862. Retrieved 5 June 2018.

Washington Post Editorial Board. 2017, July 15. Solitary Confinement Is Torture. Will the Bureau of Prisons Finally Stop Using It? *The Washington Post.* https://www.washingtonpost.com/opinions/solitary-confinement-is-torture-will-the-bureau-of-prisons-finally-stop-using-it/2017/07/15/f719de20-68c6-11e7-8eb5-cbccc2e7bfbf_story.html?utm_term=.adc722effac4. Retrieved 5 June 2018.

Part III
Ethical Assessment

Chapter 8
Ethical Theory

> *"The cell had no ceiling. It was raining. At midnight they threw something at my sister's feet. It was my brother Ayad. He was bleeding from his legs, knees and forehead. ... The next day they took away his body."* The US military later issued a death certificate, seen by the Guardian, citing the cause of death as *"cardiac arrest of unknown etiology".* The American doctor who signed the certificate did not sprint his name, and his signature is illegible.
> —Luke Harding, *The Guardian* (U.K.), 20 Sep 2004.

Overview

The last part of the book, Part III focuses on ethical assessment. In this chapter we look at the major ethical theories and how they can help provide a framework for assessing doctors' involvement in torture. This survey and application includes Teleological Ethics (which prioritizes end goals and consequences), Deontological Ethics (which prioritizes moral duty and obligations), Virtue Ethics (which prioritizes a life of virtue and the development of moral character), and Feminist Ethics (which prioritizes relationships and moral agency).

After a brief overview, each theory is applied to the use of torture and the question of doctors' participation. Applying these theories gives us a handle on whether abuse and torture can ever be justified on moral grounds. In order to arrive at an answer, we need to take into account both short-term and long-term consequences.

8.1 Introduction

To get a handle on doctors' active or passive participation in torture, it is useful to look at the major ethical theories. They are a powerful vehicle of analysis and reflection. Each ethical framework can assist us in examining theoretical, hypothetical and real-world cases. They help structure an inquiry into either conceptual or experiential concerns. The conceptual, more abstract, level is Metaethics, while the more concrete, practical level is Normative Ethics. Let's get an overview of both and see how the major ethical theories apply to the issue of torture.

In the former, Metaethics, we examine ethical theories, moral concepts, and value-laden terms such as "good" vs. "bad," "virtuous" vs. "vicious," "praiseworthy" vs. "shameful". Here's where we survey the different theories, note any differences, assess their value as theoretical models, and contrast their principles.

We also evaluate the criteria for moral agency, rationality, competence, and moral duty—and concepts like "intention" and "obligation" that have ethical force when evaluating accountability. Unpacking or constructing definitions in ethical arguments also falls under the Metaethical umbrella. Given the wrangling over terms and definitions in the torture debate, there is considerable action to be had in the Metaethical realm.

In Normative Ethics, the boots are on the ground. Here's where we apply the theories, arrive at ethical decisions, evaluate moral reasoning in context, offer advice, and make value judgments. This brings ethics into the world, where we decide what is right or wrong in a particular situation and decide if action is required.

As such, we are looking at the practical, experiential side of ethics. We consider the ethical dilemma, weigh the various factors, contemplate the options, and arrive at a decision or course of action. So, for example, when we call an interrogation "brutal" or judge a participant a "torturer" we are in the realm of Normative Ethics.

8.2 Moral Agency and Culpability

Moral agency is of fundamental importance in Normative Ethics by determining if an individual is capable of ethical decision-making. Only then can we hold them responsible for their actions. The two criteria of moral agency are free will (volition) and rationality (cognition, competence).

To be held responsible for our actions and intentions, we must be free to decide and act. This means we are not under compulsion, coercion, or threat to do one thing or another. Those acting under duress may not feel free to make a decision or act on their own volition and, consequently, might be excused from moral culpability.

The second aspect of moral agency, rationality, entails the knowledge of good and evil. To be held responsible for our decisions and actions we must have the cognitive ability to tell right from wrong. A great deal turns on competence when it comes to accountability. If the person is not competent, assigning responsibility comes into question or collapses altogether. Being a moral agent—or not, if failing to satisfy one or both criteria—must be clarified before making a judgment regarding accountability.

8.3 The Major Ethical Theories

With moral agency in mind, let us now turn to the major ethical theories. Each theory provides a structure and set of expectations and parameters for decision-making. They help illuminate the moral aspects of an ethical dilemma by shedding light on the values and boundaries that come into play. The ethical theories we will look at here are:

- Teleological Theories—focused on end goals and potential consequences (Ethical Egoism, Ethical Relativism, Utilitarianism)
- Deontological Theories—focused on moral duty and obligations (Kantian Ethics, Ross's Prima Facie duties, Rawls's Justice Theory)
- Virtue Ethics—focused on moral character and living a life of virtue (Aristotle, Confucius)
- Feminist Ethics—focused on moral agency and relationships

Let's proceed by looking at each of these theories.

8.4 Teleological Ethics

Teleological Ethics focuses on goals and consequences and not intentions or means to an end. It is often referred to as Consequentialist Ethics, given the emphasis on objectives. The overriding concern is to maximize goods or social benefits and minimize harms or disadvantages for oneself, a group, or the greater society, depending on the targeted population.

Teleological theories fall into three categories: Ethical Egoism, Ethical (or Cultural) Relativism, and Utilitarianism. They have similar emphases (namely end goals), but differ in scope, as we will see. A key similarity is that they all subscribe to what is called the *Principle of Utility,* which holds that we ought to maximize benefits or gains and minimize harm or losses. They differ in terms of whose interests should prevail and whose benefits should be maximized.

8.5 Ethical Egoism

Ethical Egoism has the narrowest scope in Teleological Ethics. It focuses solely on the individual and puts self-interest above all else. The *modus operandi* is straightforward: "Look out for Number One." Choose the option that will produce the greatest benefit and least harm for yourself—so commands the Principle of Utility for Ethical Egoism.

The interests of others matter only if you stand to gain or lose by granting their significance. Everyone else's interests recede unless it would be advantageous to factor them into the decision-making. Any concern for the welfare of others, any altruism, is not on the radar of the Ethical Egoist.

The foremost proponent of Ethical Egoism is Ayn Rand, whose philosophy has numerous adherents in politics and business. Like the capitalist who favors beating the competition even if the competitor will suffer adversity, Rand puts the individual at the center of her moral universe. As she says in *Atlas Shrugged*,

> Pride is the recognition of the fact that you are your own highest value and, like all of man's values, it has to be earned—that of any achievements open to you, the one that makes all others possible is the creation of your own character (1957, 776).

Ethical Egoists would condone such thinking.

8.6 Applying the Theory

As indicated, Ethical Egoists look out only for themselves. They show little sympathy or motivation to help others unless they stand to benefit. As a result, a medical professional would only intervene to stop torture if it was personally profitable. Coming to the aid of the victim would not be considered a necessity unless there was personal gain. As a result, an Ethical Egoist would not be compelled to help the victim of abuse without some payoff.

Of course, the long-term consequences of enabling torture may not be in the back of the mind of a participating doctor. In addition, it's not always possible to see what lies further down the road. However, inflicting harm upon another may be difficult to brush off. In fact, the evidence suggests that assisting abuse can't easily be erased. As a result, the doctor-torturer may suffer post traumatic stress or overwhelming guilt as time goes on. An Ethical Egoist would be wise to take that into account, even though looking ahead at the repercussions can seem too hypothetical to be accorded much weight.

This is particularly the case when torture or "enhanced interrogation" is encouraged or tolerated, as with "It's time to take the gloves off." Of course, there are times when medical personnel might feel coerced to go along with abuse while personally opposed to torture. Peer pressure may carry weight as well.

Even so, moral revulsion may surface and generate second thoughts. Unless it is beneficial to speak up, doctors would, nevertheless, feel it's in their best interest to assist the degradation and brutality, or stay silent. That includes falsifying records or otherwise covering up torture, as well as doing nothing to contact those higher up the ladder and report abuse. Self-interest overrules any moral qualms about such violations of human rights. Or so thinks the Ethical Egoist.

8.7 Ethical Relativism

Ethical Relativism foregrounds the interests of a specific community, group, organization, or cult. There may be shared principles and values, but the bottom line is that the group's interests take precedence over those of individual members. That "group" could be a culture, subculture, professional organization, ethnic group, religious organization, a bunch of activists or political allies, a gang, terrorist cell, or so on.

With Ethical Relativism the group's interests are to be maximized, even if the individual is on the losing end of the spectrum. The Principle of Utility in this case designates the group in question as the focal point. What benefits the group may be disadvantageous to individual members and entail personal sacrifices. That the individual accedes to this value system means leaving their independence and, to some degree, their rationality at the door.

Adopting such priorities and conforming to the group's goals and standards makes Ethical Relativists vulnerable to questionable or even dangerous decision-making. Think of suicide ("martyr") bombers who give their lives to honor the group's values and set of beliefs. That's the dark side of this theoretical framework.

On a more positive note, Ethical Relativism allows for multiple points of view and the possibility for community values and influence to steer misguided members away from a destructive or morally bankrupt mindset. From a global perspective, this could offer advantages for less powerful groups (e.g., communities, ethnic or religious minorities, and political factions) relative to those with more status and clout.

One issue with Cultural Relativism is whether the values of the group are just relative in-kind or whether you can have universal principles that some or all cultures subscribe to while allowing differences with other principles and values. The extreme view is that there is no universally shared value; therefore, values are simply relative to the various groups and cultures around the globe. Nevertheless, many Cultural Relativists believe that there can be some shared or universal values across cultures.

8.8 Applying the Theory

Ethical Relativists, by putting their group affiliation above individual interests, will generally accede to the group's decisions and policies. Even if they have doubts or disagree with a community's course of action, Ethical Relativists will not be inclined to buck the system. Because of their lower status and limited role in the decision-making, Relativists' compliance with the group's conventions and standards is pretty much a given.

As a result, a medical professional personally opposed to torture or abuse may, nevertheless, go along with it if doing so conforms with the group's wishes. In prison settings, where there is relative isolation from the wider society, health professionals may feel pressured to follow the guards who have set abuses in motion. Coming to the aid of the victim may only be deemed a necessity if the there's a shift in attitude. As Bioethicist M. Gregg Bloche (2004) notes,

> According to press reports, military doctors and nurses who examined prisoners at Abu Ghraib treated swollen genitals, prescribed painkillers, stitched wounds, and recorded evidence of the abuses going on around them ... these medical professionals had a duty to tell those in power what they saw. Instead, too often, they returned the victims of torture to the custody of their victimizers. Rather than putting a stop to torture, they tacitly abetted it, by patching up victims and staying silent.

In Rory Kennedy's documentary *Ghosts of Abu Ghraib* some of the soldiers spoke about joining their buddies in the vicious treatment of detainees. They went along with the group in crossing a moral line. Photos and detainee testimony indicate that medics also played their part in an atmosphere of brutality. Such conformity came to be viewed with chagrin by the time of Kennedy's interviews.

In the documentary *Taxi to the Dark Side*, similar sentiments are also expressed. As Sgt. Anthony Morden said of his participation at Bagram, "Sometimes I feel that I should have gone with my own morality more than what was common."

8.9 Utilitarianism

The most influential form of Teleological Ethics is Utilitarianism. It holds that we should maximize good or happiness and minimize evil or unhappiness for the greatest number of people. Aim for societal benefits; favor consequences over the intentions or moral principles behind the act.

Utilitarians put society's interests above those of any individual. This allows for individuals to be used or even sacrificed for the public good (e.g., with the death penalty, military conflicts, and non-therapeutic medical experimentation). The good of the majority takes precedence over that of a minority.

Utilitarians favor a cost-benefit analysis, where they choose the option with the highest ratio of gains. Weigh the potential consequences and then select the best of the lot. The means are of lesser value than the ends, since the objective is given prominence.

The foremost Utilitarians are Jeremy Bentham, John Stuart Mill, and G.E. Moore. They each set out their version of the Principle of Utility drawing from the following template:

> Choose that act that will result in the greatest_____ and the least_____ for the greatest number of people.

Each ethicist fills in the blanks differently. Bentham seeks to maximize pleasure and minimize pain; for Mill it's happiness vs. unhappiness (in terms of social benefits); and for Moore, good vs. evil. They do not seek to make this a universal command. Since they focus on the society rather than the individual, their goal is to arrive at the most for the most, the majority. That is why, in such an equation, it is morally permissible to forfeit individual rights as a means to that end.

There are two main forms of Utilitarianism—Act and Rule. They both seek to maximize good or happiness and minimize evil or unhappiness, but differ in scope. The one (Act Utilitarianism) has a much narrower targeted population than the other (Rule Utilitarianism), as we will see.

Act Utilitarians would have us choose what leads to the best consequences overall for the greatest number of people *directly affected* by the act in question. Weigh each case with this in mind. The cost-benefit assessment of the specific dilemma shapes the decision and, thus, Act Utilitarians proceed on a case-by-case basis. In contrast, Rule Utilitarians seek to maximize good and minimize harm for the greatest number of people, not just those directly affected. That means looking at case as a possible precedent. Rather than proceeding on a case-by-case basis, approach the ethical dilemma in terms of its wider application to the society in general.

8.10 Applying the Theory

Utilitarians seek to maximize benefits for the greatest number of people. The focus may be upon those directly involved (as with Act Utilitarianism) or across the board (as with Rule Utilitarianism). Consequently, whether a doctor should support torture depends upon short-term versus long-term benefits. In the short term, abusive interrogation techniques may bring about some positive results or useful information (so-called "actionable intelligence").

Mind you, whether that really is the case may not be readily apparent. As with terrorist suspects Khalid Sheikh Mohammed and the Tipton Three, the detainee-victim may invent "information" in order to get the torture to stop. Their "confessions" contained so many fabrications as to be ludicrous. And worthless—as with the Tipton Three's false "admission" of working for Osama bin Laden when they were actually employed in Britain. Consider, also, the remarks attributed to high-profile detainee Abu Zubaydah:

> After being subjected to marathon interrogation and torture sessions, Abu Zubaydah said he asked his interrogators, "Tell me what you want me to say, I will say!" Sometimes, he acknowledged, he "just said things that were false and that I had no basis to know or believe, simply to get relief from the pain" (Leopold 2016).

However, there may be potential gains thought to override the harm inflicted. If so, torture gets a green light. Utilitarian doctors would sacrifice the well-being of the prisoner or detainee for that presumably higher goal. Torture would thus be seen as a means to an end, a tool whose value legitimizes its use.

On the other hand, and here's the rub, the long-term consequences may diminish or negate the benefits of torture. For example, torture may bring about retaliatory actions that escalate the conflict. It may damage relationships between the various parties or be a political nightmare.

As Professor of Medicine Steven H. Miles points out, it was only after the scandal at Abu Ghraib that the beheadings of Americans and allies began. "Before the [Abu Ghraib] photos became public," he observes, "every POW returned alive, but not afterward. [TV carried the Abu Ghraib photographs on April 29, 2004. The first of 11 beheadings in Iraq occurred 12 days later]" (Thieme 2016). The brutality and torture at Abu Ghraib set in motion an escalation of tensions and moral outrage, thus complicating an already volatile situation.

That being the case, health professionals who are Utilitarians are caught on the horns of a dilemma. Since long-term consequences are not as easy to predict as what can be accomplished in the short-term, Utilitarians generally go for the short-term. They would then tolerate torture in the belief that it furthers the end goals and brings about the desired result. If some innocent people suffer in the bargain, so be it. The objectives—the ends—justify the brutality, particularly given the high stakes in the war on terror.

Doctors who go down this path may persuade themselves that the short-term benefits compensate for any long-term difficulties. That those difficulties include a violation of human rights and the transgression of professional codes tend not to be given much credence. However, problems will most certainly bubble up to the surface as time goes on, making it hard to repair the damage done. As a result, Utilitarian doctors would be advised to downplay any short-term gains that torture may provide and give more attention to what the long-term may bring.

A Rule Utilitarian would advise the health professional to choose that act that would result in the best consequences for the greatest number of people, the majority. This asks us to consider each case as a precedent, thus generalizing the decision ("rule") and applying it to all similar cases.

This is quite different from Act Utilitarianism's focus on the particular situation, its context, and only the people directly affected by the decision. In the case of torture, Rule Utilitarianism broadens the playing field to the level of generating policy. Contrary to public denials, torture became systemic and incorporated under the guise of "enhanced interrogation".

Should torture be permissible on a policy level and, therefore, be used on a much larger scale? Is the widespread use of torture to be tolerated as a legitimate means of handling recalcitrant suspects or arriving at presumably useful information that can further the society's goals? A doctor who is a Rule Utilitarian would need to answer these questions before giving a stamp of credibility to the question of torture.

8.11 Deontological Ethics

Deontological Ethics emphasizes duties and intentions rather than end goals. In that sense, this theory is the flip side of Teleological Ethics. The focus here is on moral obligations, not consequences. Ethical decisions are approached in terms of principles, ethical codes and moral duties; as with "Don't harm living creatures," "Act with integrity," "Never give a vegan ice cream," and so on. In contrast to Utilitarians, they believe that intentions inform actions and undesirable consequences should not determine whether a decision is good. Moral obligations rule.

What follows is that Deontological Ethicists focus on human rights, not societal benefits. Their emphasis is on the individual and not the society. This is the case, even if the society as a whole would benefit more from sacrificing a minority for the good of the majority. Using one for the good of all is morally unacceptable. So, for example, Deontologists deplore the Tuskegee syphilis study and the human radiation experiments. And, so, torture would be impermissible. Whatever beneficial knowledge was gained did not warrant the exploitation of innocent, uninformed subjects.

8.12 Kantian Ethics

Immanuel Kant is Numero Uno in Deontological Ethics. His main concern is the individual and respect for persons. One way this is accomplished is universalizing ethical decision-making.

In Kant's view, human dignity is identified with the capacity for rationality and, therefore, he puts a lot of importance on moral agency. Moral agents must be rational and have free will. Those having both characteristics, Kant argues, have equal moral status. Excluded from the set of moral agents are children, incompetent adults, and anyone else whose mental capacity is such that they cannot make a reasoned decision.

At the center of the Kantian universe are two ethical principles he considers moral imperatives—commands. They function as categorical principles, because they apply to all rational adults (= moral agents). Moreover, these two commands are meant to guide our ethical decision-making; consequently, they carry considerable moral weight.

Kant's Two Moral Imperatives

The Categorical Imperative
Act in such a way that you would have it become a universal law.

→ Universalize ethical decision-making.

The Humanitarian Principle
Always treat others as an end in themselves, never merely as a means.

→ Treat others with respect and human dignity.

Kant's first imperative aims at universalizing moral reasoning. Before settling on a decision, ask, "What if everyone did this?" If it would be acceptable for all to follow suit, then the action is morally permissible. Otherwise not. Only if it's okay for everyone to make the same choice ought I proceed: I can't be an exception to the rule. Instead, I should strive to be a moral role model. That is the thrust of his Categorical Imperative.

Kant's second formulation of the Categorical Imperative is the Humanitarian Principle. It directs us to treat people as ends in themselves. Never use others merely as a means; rather, treat people with dignity and respect and, subsequently, honor their human rights.

For Kant, there are no exceptions to his two principles, even if the consequences of following them may be disastrous. By prioritizing moral duty over end goals, the results may create a wasp's nest of problems. Serious problems. This is further complicated by Kant's emphasis on the Principle of Veracity to always be honest. But what if telling the truth to the Hillside Strangler would reveal the location of his next victim—your mother? Kant would say we are obligated to be honest no matter what. No exceptions. Clearly this is problematic and not a path we'd always want to go down.

Because of this drawback—following duties and obligations that could unleash horrific consequences—some Kantians have modified the theory. Why not allow for an escape hatch if things go South? One such option is Philosopher Sissela Bok's Principle of Non-maleficence. This introduces a way to block disasters. Here's her recommendation: The Principle of Non-maleficence has priority over the Principle Of Veracity. So if telling the truth would result in grave harm, we should either stay silent or tell a lie. Avoid harming someone because of a rigid (and possibly foolhardy) commitment to honesty and the Categorical Imperative.

8.13 Applying the Theory

Kantians put moral duty and intention above potential consequences and value individual rights over social benefits. As a result, they would be reluctant to support torture or abuse. Their advice to doctors would be to refuse to participate; instead take steps to assist the victims wherever possible. And blow the whistle. Complicity is not an option.

The Humanitarian Principle makes it clear that people should not be used merely as a means to an end, as in the case of torture. Additionally, the Categorical Imperative calls us to act in a way that we would willingly have others follow. Both principles send an unequivocal guideline to medical professionals not to violate human rights. In addition, uphold your fiduciary duties to your patients, regardless of the context in which you find them.

Opening the door on torture is not something a Kantian would make into a universal law. Hardly. Instead, "Do no harm," is the moral command to follow when seeking actionable intelligence. Medical professionals who are Kantians would have nothing to do with the use of torture. There's no way the respect for human dignity is consistent with the abuse of another. Brutality can't possibly be made into a universal law. As a result, doctors contemplating torture would get a clear message from Kantians: Under no circumstances should you go this route.

8.14 W.D. Ross's Prima Facie Duties

Deontological ethicist W.D. Ross agrees with Kant on the importance of moral duty. However, he thinks Kant was too fixated on the present. Factoring in the past and future, he contends, will provide a more balanced approach to ethical decision-making. We may, for instance, feel obligated to make reparations for harms done in the past or make beneficial policies for future generations.

Ross sets out what he calls "Prima Facie" duties. On the surface they are of equal value. Only when put into play do some come to the foreground, leaving the others with lesser significance (if any) on the sidelines. On a case-by-case basis, the most relevant duties will likely be apparent. That should be intuitively obvious thinks Ross. Here are his seven moral duties:

Ross's Prima Facie Duties

1. Honesty and promise-keeping (Veracity)
2. Beneficence (Do good)
3. Non-maleficence (Do no harm)
4. Justice
5. Gratitude (Loyalty)
6. Reparations (Compensate for harm done)
7. Self-improvement

These duties are in no particular order; they are of equal merit until we face a particular ethical dilemma. At that point we would look at the context to determine which of the seven duties apply. For example, Ross values honesty, but justice or non-maleficence may take precedence if telling the truth would result in serious harm. He thinks we will intuitively know which duties should guide us and which ones have little or no relevance.

8.15 Applying the Theory

Followers of Ross address ethical decision-making by drawing from the applicable prima facie duties. They would consider torture morally abhorrent. The duties of non-maleficence, justice, and beneficence would direct the medical professional to actively oppose the use of torture in virtually all circumstances. Also, the duty of self-improvement would prohibit medical personnel from ceasing their professional duties to their patients. It could hardly make the doctor a better person to play any role in inflicting torture on another, regardless of their guilt or innocence.

Professional associations like the WMA, AMA, and most others concerned with the medical profession take strong stands opposing the use of torture. As a result, doctors' professional standing as a factor of self-improvement would require them to reject the use of torture and to speak out against its use.

World Medical Association president James Appleyard calls on physicians to step up to the plate. "Doctors cannot turn a blind eye to what is going on," he contends. "Torturers rely on the cloak of secrecy. The ability to expose acts of torture is crucial to its prevention" (World Medical Association 2004). Bioethicists like Jonathan H. Marks have suggested this be mandated by the professional organizations with sanctions placed upon those who violate them.

8.16 John Rawls's Justice Theory

John Rawls is another Deontological Ethicist who emphasizes moral duty. Like Kant, he focuses on the individual and seeks to universalize ethical decision making. Like Ross, he thinks justice is an important moral duty. He is particularly interested in addressing injustice within social institutions.

Rawls's goal is to eliminate personal bias so we distance ourselves from our personal attachments and affiliations (nationality, religion, political party, race, gender, etc.). He believes this would level the playing field and put in place a system of justice as fairness. We could then arrive at a social contract that would eliminate prejudice and other forms of injustice and get that much closer to universalizing human rights.

To accomplish this goal he recommends we adopt a "veil of ignorance" when constructing policies or making decisions about who should do what. We need to separate ourselves from any vested interests so we eliminate sources of bias. As much as possible we should extricate forms of prejudice from how we think—and behave.

Rawls sets out three principles aimed at helping us construct a society rooted in justice and fairness. They are as follows:

Rawls's Three Principles

Principle of Equal Liberty:
Each person is to have an equal right to the most extensive system of basic liberties compatible with a similar system of liberty for all.

→ Universalize human rights.

Principle of Equality of Fair Opportunity:
Offices and positions are to be open to all under conditions of equality of fair opportunity so that persons with similar abilities and skills should have equal access to offices and positions.

→ Provide equal opportunities.

Difference Principle:
Social and economic institutions are to be arranged to maximally benefit the worst off.

→ Favor the least advantaged

8.17 Applying the Theory

Followers of Rawls seek a society free of prejudice and bias, with institutions championing fairness for all. Like Kant, they strive toward a universal system of human rights, where moral agents have equal standing. Honoring those rights is of paramount importance.

As such, they would oppose the use of torture. Rawls's first principle, the Principle of Equal Liberty, is particularly relevant in cases involving torture. The Geneva Conventions' prohibition against torture embodies the values expressed in Rawls's principle. No amount of manipulating language should blind doctors from separating right from wrong.

The notion that the Geneva Conventions' torture prohibition is only meant to protect prisoners—not detainees—rests on linguistic quicksand. The assertion is that prisoners have been charged with a crime, whereas detainees are neither charged nor convicted, so the Geneva Conventions simply do not apply. Thus, they lack the protections available to prisoners. Doctors would be advised not to fall for such reasoning.

Even if detainees were not actually covered by the Geneva Conventions does not mean that we should dispense with the moral imperatives or the liberty rights recognized by other codes and principles acknowledged by governments around the globe. Human rights do not exclude those held in detention centers who may go the rest of their lives without facing any charges or having access to legal counsel or the evidence against them. Doctors should not play any role in keeping such unjust policies in place.

Those advocating "enhanced interrogation" should seek alternatives that protect respect for individual dignity and minimize bias. Like other Deontological Ethicists, adherents of Rawls's justice theory reject prioritizing ends over means. Simply put: the maximization of societal benefits is trumped by the preservation of human rights. Doctors take note. However worthy torture may seem, it has no place in a just society.

Furthermore, "the veil of ignorance" would have us to tear down any "Us versus Them" mentality. Rawls would find such divisive thinking to be abhorrent. Consequently, he would advise medical personnel to use terms like "enemy combatant" with care, if at all, and steer clear of any complicity with torture or brutality.

Rawls would agree with Kant that honesty is a significant obligation in this situation. Falsifying documents and covering up abuse are unacceptable. He would commend doctors who step forward and disclose any use of torture. And for those who enable torture, "Turn yourself in," as Bloche (2004) would advise. Rawls would second that opinion and agree with Ross that justice is an important moral duty that doctors should not lose sight of. Torture should be out of bounds for all medical personnel, regardless of any pressure to show their loyalty to the fight against terrorism.

8.18 Virtue Ethics

Virtue Ethics is so named because of its focus on virtues and vices. It sees virtues as paving the way to the good life, a life of flourishing and well-being. In contrast to Teleological (goal-based) Ethics and Deontological (duty-based) Ethics, Virtue Ethics considers moral character the primary ethical concern. Living a virtuous life free of vices is the ideal.

Aristotle is the heart and soul of Virtue Ethics. That said, he is not alone. Confucius also focuses on leading a virtuous life. He emphasizes the particular roles we play in our relationships, such as being a son or daughter or roles tied to one's profession. I will focus here on Aristotle, but it is worthwhile to look into Confucius as well.

Aristotle looks at rational individuals (= moral agents) and asks us to examine our moral character. Put that under the spotlight. Neither consequences nor moral obligations take priority over a life of virtue. That is the path to finding purpose in our lives. Aristotle emphasizes both intellectual virtues and moral virtues. His intellectual virtues are as follows:

Aristotle's Intellectual Virtues

1. Artistic knowledge and creativity
2. Analytical and logical knowledge
3. Practical wisdom ("street smarts" about what we ought to do)
4. Philosophical wisdom (about ultimate things)
5. Understanding and comprehension (for making judgments)

Most people focus on Aristotle's moral virtues, but intellectual virtues should not be discounted. His moral virtues are as follows:

Aristotle's Moral Virtues

1. Courage
2. Self-Control
3. Generosity
4. Magnanimity
5. Pride
6. Even Temperament
7. Honor
8. Integrity
9. Wit
10. Friendliness
11. Modesty
12. Righteous Indignation

To get there from here we need to practice a life of moderation. Seek the mean between the extremes of too little (deficient) and too much (excess). Of course, moderation may not be always the best choice. In the face of injustice or acts of violence against innocent others, the morally correct response might be what Aristotle called "righteous indignation"—what we might call moral outrage.

Generally, however, Aristotle recommends the middle path. Keep in mind the moral spectrum of deficient → mean → excessive. For example, courage is the mean between cowardice and recklessness, and generosity is the mean between being stingy and being wasteful. It's not always easy to steer clear of the extremes, but we should strive to do so to achieve moral excellence.

8.19 Applying the Theory

Virtue Ethicists would tell doctors to focus on their moral character. Seek a life of excellence by cultivating the moral virtues and avoiding the vices. If we look at the torture debate, Aristotle's advice would be, "Don't even think about taking part in the abuse of others." The virtues of courage, self control, and integrity should steer you in the right direction.

That means not simply refusing to participate; it also means standing up to those who would have you take part. Being virtuous is not always easy, but succumbing to vice will never build moral character.

Caring for patients and using one's talents and training to help others is a vital part of the health profession. Doctors should not turn away from the fiduciary duties that accompany those talents and training. Under no circumstances would Aristotle see torture as a legitimate option a doctor should consider.

Looking over Aristotle's moral virtues, it seems clear that torture would be unacceptable. Medical personnel would be ill-advised to play any role in the mistreatment of another person. Aristotle's overarching virtue of justice would prohibit acts of torture, regardless of the motivation or justification. Furthermore, in the face of torture, the most appropriate response would be righteous indignation, not complicity. Unfortunately, doctors in the field have not expressed nearly as much righteous indignation as is called for. That can, and should, change.

8.20 Feminist Ethics

Feminist Ethicists view relationships as a central force in shaping moral reasoning. In contrast to Deontological and Teleological theorists, they do not consider moral duty or consequences to be at center stage. In assessing values and actions by factoring in relationships, they tend to the practical and experiential dimension of life. They contend that "human agents are not fundamentally single-minded, rational, self-interested choice-makers but social beings whose selfhood is constituted and maintained within overlapping relationships and communities." (Donchin 2004/2015).

Philosopher Mary Anne Warren takes issue with the importance ethicists such as Kant place on rationality. The price in overlooking or downgrading the role of relationships is seen as too high. Ethicist Virginia L. Warren agrees. In her estimation,

> The dominant trend in philosophical ethics has been to regard people as best able to decide what is moral when least tied to place and time, when least connected through ties of partiality to family and community. Ideal moral decision-makers are viewed as common denominators—e.g., rational egos (Kant) or calculators of utility—who are more likely to adopt the proper universal perspective when the veneer of particularity is stripped away.
>
> [However] although some moral agents may adopt the common denominator moral perspective without feeling that anything of value is lost, others may feel the loss intensely. The reason for this loss is that persons whose unique experiences have been largely omitted from the dominant culture—e.g., women, Blacks, gay males, and lesbians—may find the stripping away of particularity from the moral observer to be anathema to self (1992, 33).

Feminist Ethicists take issue with such thinking. We are not neutral beings, hollowed out by casting aside our unique, and human, characteristics. For that reason, Mary Anne Warren considers relational properties important when developing a moral theory. These are properties necessarily involving more than one person—like being a daughter, friend, or naturalized citizen. She sees relationships as integral to moral status and, thus, to moral reasoning. Because of this theoretical shift Feminist Ethics is often called Care Ethics, or Ethics of Care. Rita Manning's two elements demonstrate why this is the case:

Manning's Two Elements to an Ethics of Care

A Disposition to Care
We are called to attend to others' needs.
Caring is a goal, an ideal.
→ People are disposed to care about one another.

The Obligation to Care For
Caring for another requires action, not just intentions.
Caring for others can be expanded to communities, values, or objects.
→ Responding to the needs of others is a moral obligation.

Caring for others highlights the significance of relationships, but it is not the only value. Manning argues that we should weigh both care and justice when confronted with ethical choices. That means paying attention to rules and rights but not to the exclusion of relationships. In other words, neither justice nor care should be neglected. We should "conceive of ethics as a social institution whose chief function should be to justly promote the well-being of all," Virginia Warren observes (1992, 33).

Feminist Ethicists attempt to address the shortcomings of Deontology and Utilitarianism by taking ethics off its pedestal and into the context of our lives. This makes it less abstract or theoretical and more concrete, practical. It's not that principles don't matter; it's just that other concerns count as much or more when plotting a moral course. Feminist ethicists recognize this concern by recommending a focus on both care and power in ethical decision-making.

Rosemarie Tong emphasizes this point when she says,

Whereas a care–focused feminist approach to ethics emphasizes values that have been culturally associated with women (e.g., nurturance, care, compassion), a power-focused feminist approach to ethics emphasizes the need to eliminate those social, economic, political, and cultural systems and structures that maintain patriarchal domination and work against the establishment of a gender-equitous world (1997, 48).

In short, Feminist Ethics follows neither Deontology nor Teleological Ethics. Neither moral duty nor end goals should be definitive in ethical decision-making. Instead we should factor in our disposition to care without losing sight of power dynamics and working for justice.

8.21 Applying the Theory

As Rosemarie Tong notes, Feminist Ethics takes two major approaches to an ethical dilemma. Both apply to the torture debate and the quandary a health professional may face. First, the Ethics of Care would advise doctors to honor their doctor-patient duties and think long and hard before casting them aside. The relational properties of being a medical caregiver—and that means *care* giver – should factor into their ethical decision-making. They should not just function with abstract principles operating at a distance from the lives they or their patients lead.

Secondly, Feminist Ethicists would have doctors pay close attention to power-relationships and guard against those that further oppression, particularly of vulnerable groups. Certainly those held in prisons and detention facilities have a lower moral status than those putting abusive structures in place. Medical personnel should reject playing any part in such oppression.

Detainee lawyer Alka Pradhan raises serious questions about power and oppression. As reported by Jeffrey E. Stern, she "sees a system that is deliberately and aggressively racist. 'This is a legal venue that was designed for noncitizen Muslim males, right?'" Furthermore, "If we had white guys from, you know, France being held, I remain convinced they would not have been tortured, I don't think," she says. "I don't think they would have founded an entirely different legal system for them. I don't care what you think they did" (Stern 2017). Doctors should take such concerns to heart and address the extent to which racism and other forms of prejudice factor into the maltreatment of prisoners and detainees.

Short of actually killing the other person, surely the worst type of oppression is torture. To accede to such behavior will take doctors down a road at odds with their professional standards. By disregarding the values of the healing profession, there will be a steep price to pay for all concerned.

We see this with the torture of detainee Abu Zubaydah. He testified that, "In between waterboarding sessions, he was placed in what he called a "dog box," a wooden container that was about 2½ feet long, 2½ feet wide and 2½ feet high." He said, "The pain in the small box was unbearable. I was hunched over in a contorted way and my back and knees were in excruciating pain" (ProPublica 2016). Where were the voices of opposition?

Simply put: Doctors should not stay silent in the face of oppression. Although there may be some contexts in which the duty to care for the patient could be rescinded, they do not include those where people are being tortured.

The very fact most detainees have no access to information about their "crimes" indicate cracks in the wall of justice. With indefinite detention in place detainees are stuck in a Kafkaesque limbo. Decades could transpire without learning the charges against them. This alone is a red flag that doctors should not ignore. Feminist Ethicists would find this situation deplorable and urge medical personnel to register their opposition.

8.22 Conclusion

As we can see from the survey of the major ethical theories, most ethicists would advise medical personnel to have nothing to do with the torture of another person. And the thought that any would falsify death certificates, cover up abuse, or commit other acts of dishonesty would be viewed with alarm.

Brutality casts a long shadow. Participants may not foresee the price they will pay as time goes by. The only ethical theories that would consider torture an option are Ethical Egoism and Ethical Relativism. In the first case, Ethical Egoists would likely participate if there was a payoff. If they think it would benefit them, any doubts would be cast aside, at least temporarily. However, short-term gains may be offset by long-term disadvantages. And not being able to predict how severe are those disadvantages may be reason enough for steering clear of maltreatment and torture.

In the case of Ethical Relativists, the group they are affiliated with may justify torture by any number of political or other arguments. A doctor's professional standing would not be strong enough to override the group's decision to proceed with the abuse. However, as in the case of Ethical Egoism, the long-term harms may come back to haunt the group and any individuals associated with it.

Look, for example, at the My Lai massacre in the Vietnam War, where peer pressure or the chain of command did not negate individual responsibility for taking part. Similarly, health professionals who try to blame the group may find themselves out on the limb with no rescue in sight. In other words, Ethical Relativists cannot count on any group affiliation to save them when the bell tolls and they are held accountable for what they did.

Works Cited

Aristotle. *Poetics*. The Internet Classics Archive. http://classics.mit.edu/Aristotle/poetics.html. Retrieved 5 June 2018.
Bloche, M. Gregg. 2004, June 10. Physician, Turn Thyself in. *The New York Times*. https://www.nytimes.com/2004/06/10/opinion/after-abu-ghraib-physician-turn-thyself-in.html, Retrieved 5 June 2018

Donchin, Anne. 2015, December 16. Feminist Bioethics. *Stanford Library*. https://stanford.library.sydney.edu.au/entries/feminist-bioethics/#CarEth. Retrieved 5 June 2018.

Ghosts of Abu Ghraib. 2007. DVD. Rory Kennedy director.

Leopold, Jason. 2016, July 8. This Is How the CIA's First Captive after 9/11 Described His Years of Torture. *Vice News.* https://news.vice.com/article/abu-zubaydah-al-qaeda-describes-cia-torture. Retrieved 5 June 2018.

Manning, Rita C. 1992. *Speaking From the Heart.* Lanham, MD: Rowman & Littlefield.

ProPublica. 2016, July 8. The Terror Suspect Who Had Nothing To Give. https://www.propublica.org/article/the-terror-suspect-who-had-nothing-to-give. Retrieved 3 December 2017.

Rand, Ayn. *Atlas Shrugged.* https://archive.org/download/AtlasShrugged/atlas%20shrugged.pdf. Retrieved 3 December 2017.

Stern, Jeffrey E. 2017, December 19. Alka Pradhan v. Gitmo. *The New York Times* magazine. https://www.nytimes.com/2017/12/19/magazine/alka-pradhan-v-gitmo.html. Retrieved 3 December 2017.

Taxi to the Dark Side. 2007. Script. https://www.springfieldspringfield.co.uk/movie_script.php?movie=taxi-to-the-dark-side. Retrieved 3 December 2017.

Thieme, Richard. 2016, January 13. Interview with Steven Miles: The Torture-Endangered Society. *ThiemeWorks.* www.thiemeworks.com/write/archives/steven_miles_interview.htm. Retrieved 3 December 2017.

Tong, Rosemarie. 1993. *Feminine and Feminist Ethics.* Belmont, CA: Wadsworth Publishing.

———. 1997. *Feminist Approaches to Bioethics.* Boulder, CO: Westview Press.

Warren, Mary Anne. 1997. *Moral Status.* Oxford: Oxford University Press.

Warren, Virginia L. 1992. Feminist Directions in Medical Ethics. In *Feminist Perspectives in Medical Ethics*, ed. Helen Berquaert Holmes and Laura M. Purdy. Bloomington: Indiana University Press.

World Medical Association. 2004, June 12. Physicians Should Report Acts of Torture, says World Medical Association President. *Medical News Today.* www.wma.net/e/press/2004_14.htm. Retrieved 3 December 2017.

Chapter 9
Applied Ethics: Principles and Perspectives

> *Essentially the doctors and psychologists were built into the entire torture system. They weren't simply bystanders who were called in to respond when the system went off the rails.*
> —Steven H. Miles.

Overview
Here in the last chapter of the book, I focus on two applied ethical frameworks and key professional and international guidelines prohibiting torture. We start with two of the major figures in Bioethics, Tom L. Beauchamp and James F. Childress. Their four principles are of central importance in Western Biomedical Ethics. They are: Autonomy, Beneficence, Non-Maleficence, and Justice. Each one is discussed and applied to doctors and torture.

The second ethical framework considered is that of Ethicist Bernard Gert (2014) and the ten duties he considers universally applicable. All are then applied the question of doctors' involvement in torture—showing that doing so would be morally remiss.

We then look at professional organizations and international codes; all of which prohibit torture. These include the Hippocratic Oath, World Medical Association and its Declaration of Tokyo, the American Medical Association, the International Council of Nurses, the Istanbul Protocol, and Physicians for Human Rights. These organizations and their guidelines emphasize the responsibility and accountability of doctors taking the right path—one to "Do no harm."

9.1 Introduction

Just about everybody knows the principle "Do no harm" from the Hippocratic Oath. The moral command behind the phrase should not be taken lightly. We may want to do good and come to the aid of those facing mental or physical challenges, but it is not always possible. Nevertheless, at least we should not harm them. This sounds like something everyone should agree upon and yet here we are looking at torture.

The "Do no harm" imperative has lost much of its force when it comes to the active or passive participation of the medical profession in torture. Evidence is by now well established that doctors have played their part. Bioethicist Steven H. Miles notes three roles doctors play, namely,

- Number one, they design methods of torture that do not leave scars. For example, the so-called "rectal feeding" which is actually a medieval technique in which the intestines are inflated with a viscous material to cause intestinal pain.
- Doctors are also involved in making sure that the prisoners who weren't supposed to die didn't die.
- The third thing doctors do is they falsify medical records and death certificates to conceal the injuries of torture (as noted by Beck 2014).

Each one of these roles raises serious concerns. First, not leaving scars is just a way to hide what is transpiring, which adds to the moral lapse on the part of the doctor. It also indicates knowledge of what is taking place, so there is no question that the doctor knows an ethical line has been crossed. Awareness of guilt is thereby established.

Secondly, enabling torture by ensuring the victim does not transpire is horrifying. It is reprehensible to think that those in the healing profession should apply their talents so victims can live another day to face abuse. It's like a scene out of a horror movie or an episode on *Criminal Minds*.

Thirdly, committing acts of deception such as falsifying documents and death certificates in order to conceal torture reveals the extent to which doctors have abdicated their role as healers. In so doing they have lost their hold on personal integrity and professionalism.

As a result, we need to arrive at some guidelines that doctors will take seriously. We will look at the recommendations and moral principles of ethicists Tom L. Beauchamp and James F. Childress, and, secondly, Bernard Gert, as well as the key guidelines of the World Medical Association and the Declaration of Tokyo, the American Medical Association, the International Council of Nurses, the Istanbul Protocol and, lastly, Physicians for Human Rights.

9.2 Beauchamp's and Childress's Basic Principles of Western Bioethics

In 1979 Tom L. Beauchamp and James F. Childress set out four "basic principles of Biomedical Ethics." They are autonomy, beneficence, non-maleficence, and justice. Their focus on these four had—and still has—considerable impact in highlighting the major moral duties and cornerstones of ethical decision-making, particularly in Medical Ethics.

All four of these principles underscore the importance of human rights in the health care setting. And they certainly spotlight the wrongfulness of torture and abuse. We'll look at each of the four. The first principle is patient autonomy. It is of fundamental significance in Western bioethics and the law. It is often held up as an ethical centerpiece.

9.2.1 Principle #1: Patient Autonomy

Beauchamp's and Childress's first principle of Biomedical Ethics is autonomy. A key ruling in the legal history of informed consent *is Schloendorff v. Society of New York Hospital* (1914) establishing the right of self-determination. The New York Appeals court ruled: "Every human being of adult years and sound mind has as a right to determine what shall be done with his own body." This decision put patient autonomy on the map. It has continued to be a mainstay in court decisions regarding informed consent.

In respecting this principle, doctors have a duty to honor the right of self-determination on the part of competent adults. That includes the right to refuse treatment and the right to withdraw from medical treatment or experimentation. These rights have been violated many times over in the war on terror, indicating the need for doctors to reassess their actions and pay heed to ethical and legal boundaries.

Victims and survivors of torture have had their autonomy stripped away and are, thus, at the mercy of others—interrogators, guards, and, sadly, health professionals. This includes being subjected to environmental and dietary manipulation, beatings, sexual assaults, and medical experimentation. Their autonomy and right to informed consent are not given the credence they deserve. All competent patients have the right to refuse treatment and that includes the right to refuse such degrading or humiliating practices as forced nudity, force-feeding, "rectal hydration," and prolonged diapering.

The remaining three basic principles are also important to highlight and honor.

9.2.2 Principles #2 and #3: Beneficence and Non-maleficence

Whenever possible, Beauchamp and Childress assert, we should strive to do good. That duty is a call for moral excellence and, for doctors, it means to do what they can to help their patients. Obviously, torture has no benefits for the victim. Some hold the view that—if we assume the truth can be extracted given the application of enough pain or mental agony—torture can have positive results (other than for the victim). However, the assumption rests on awfully shaky ground and, thus, cannot be considered dependable. Or desirable.

The duty of non-maleficence is in accordance with the Hippocratic Oath's "Do no harm." That sets a minimal standard of behavior: Whatever you do for your patients, at least do not harm them. The duty of beneficence sets a higher standard by calling on medical personnel to do good and seek to maximize benefits for their patients. As much as possible prioritize the well-being of the patient.

Non-maleficence and beneficence are companion principles: If possible help others, do what you can to make an affirmative difference in the situations you face and in addressing patient needs. But if you can't reach that standard, for whatever reason, minimally do no harm.

The message is clear. Doctors who find themselves being expected to brutalize and torture others or assist in some form of mistreatment would be advised by Beauchamp and Childress to adhere to both of these duties. Keep the duty of nonmaleficence before you at all times and whenever possible hold tight to the duty of beneficence and do what you can to render help.

That means standing up to torture and intervening wherever possible and, minimally, reporting violations of human rights whenever you see them. By no means should you be complicit in abuse. However important is the war against terror, there are lines that should not be crossed.

9.2.3 Principle #4: Justice

Beauchamp's and Childress's fourth principle is justice. They are in line with Aristotle, who holds this to be the overarching moral duty—one with vast significance. What follows from this? Well, justice entails fairness and a defensible consideration of all concerned. Philosopher Wayne P. Pomerleau offers this interpretation: "Aristotle says justice consists in what is lawful and fair, with fairness involving equitable distributions and the correction of what is inequitable" (2018).

There are three aspects to this definition; namely what is fair, what is lawful, and what is equitable. For Philosopher of Law John Rawls that would call for a framework of justice that would universalize human rights and provide a system of liberty accessible for all. As much as possible eliminate bias and prejudice to maximally help the most disadvantaged.

9.3 Torture Is Unjust

This understanding of justice has repercussions for doctors and torture in terms of duties they should uphold. Steven H. Miles points out that,

> There are a set of professional codes, that are endorsed by the American Medical Association and the World Medical Association, that describe doctor's duties not only to avoid participating in torture, directly or indirectly, but also a duty to document it and to report it, going outside the chain of command if necessary (Beck 2014).

What Miles expects of medical personnel needs to be emphasized: By no means should they enable torture by falsifying records or simply failing to report and document it. No "dual loyalty" justifies such dishonesty. It may not be easy to take on the chain of command, but it is vital to pursue all possible steps to stand up against torture.

Let us now turn to Bernard Gert and his list of 10 universal duties.

9.4 Bernard Gert's Universal Moral Rules

Ethicist Bernard Gert sets out 10 moral duties of universal applicability. He argues that, "Although these rules are subject to some variation in interpretation, they can be understood by all those held responsible for their actions" (2014). They are as follows:

Gert's Universal Moral Duties

1. Do not kill.
2. Do not deceive.
3. Do not cause pain.
4. Keep your promises.
5. Do not disable.
6. Do not cheat.
7. Do not deprive of freedom.
8. Obey the law.
9. Do not deprive of pleasure.
10. Do your duty.

As with other Deontological Ethicists, Gert emphasizes the importance of moral rules in ethical decision-making. Any violation of these duties requires strong justification. He clarifies how to interpret the rules by noting:

- The rule prohibiting causing pain prohibits causing not only physical pain but also mental pain;
- The rule prohibiting causing disabilities prohibits causing not only physical disabilities but also mental and volitional disabilities;
- The rule prohibiting depriving of freedom includes prohibiting depriving of opportunities and resources;
- The rule prohibiting depriving of pleasure includes prohibiting depriving of future pleasure as well as present pleasure;
- The rule prohibiting deceiving can be broken by withholding information as well as by lying, and the rule requiring one to keep one's promises requires keeping informal agreements as well as formal contracts;
- Finally, "duty," in the rule requiring one to do one's duty, is meant in its everyday sense, where duties are determined by one's social role, job or profession, or by special circumstances. Someone who takes a job is often told what duties are involved. "Duty" is not used as philosophers customarily misuse it, to mean whatever one morally ought to do (Gert 2014).

Gert asserts that, "We are morally required to obey all of the moral rules unless we have an adequate justification for not doing so" (2014). With these ten rules as ethical boundaries, he provides a framework for decision-making.

Looking over the list of rules, virtually all of them apply to doctors enabling torture and why that is immoral. Let's see how.

9.4.1 Do Not Kill

Doctors and nurses may feel torn by their dual loyalties. They may be under pressure to show their patriotism or affiliation with those in power or in a particular group. That they may be expected to torture or even kill the perceived wrongdoers ("enemies") goes with the territory. Medical personnel need to prioritize the duty not to kill and, therefore, resist any role in killing without the strongest justification, such as self-defense.

9.4.2 Do Not Deceive

Doctors have falsified death certificates and covered up misdeeds to hide moral breaches and abuse. They may have willingly done so or were under coercion; for instance, as part of the interrogation process. They may also have subjected detainees and prisoners—their patients—to human experimentation without informing them or obtaining consent.

That's not all. *The Economist* reports that "forced injections with unknown substances" were used on unsuspecting subjects (2004). Robert Beckhusen of *Wired* (2012) echoes this concern. He asserts that prisoners/detainees were even subjected to mind-altering drugs—which calls to mind the CIA-run project MKUltra which experimented with the use of LSD on unsuspecting subjects from 1953–1973.

> Prisoners inside the U.S. military's detention center at Guantanamo Bay were forcibly given "mind altering drugs," including being injected with a powerful anti-psychotic sedative used in psychiatric hospitals. Prisoners were often not told what medications they received, and were tricked into believing routine flu shots were truth serums.

As Beckhusen (2012) points out, "It's a serious violation of medical ethics, made worse by the fact that the military continued to interrogate prisoners while they were doped on psychoactive chemicals." Gert would criticize such acts of deception. And rightfully so.

9.4.3 Do Not Cause Pain

Torture can involve mental or physical suffering, or both. Doctors cross the "Do no harm" line in assisting or initiating either of these. The impetus to focus more on torturous acts that leave no marks (such as solitary confinement, forced standing, hooding, sleep deprivation, or mock executions) should not be pursued or supported. *The Atlantic* (2014) examined the role doctors have played, noting:

> The [U.S.] Senate released its report on the CIA's interrogation program on Tuesday, revealing horrendous details of the torture tactics used on prisoners, including waterboarding, sleep deprivation, and "rectal feeding." Complicit in this treatment were several "medical officers" (it's not explicitly stated whether they hold M.D.s), who enabled, oversaw, and designed many of the techniques (Beck 2014).

9.4.4 Keep Your Promises

An implicit promise between doctors and their patients is to give appropriate and timely care. This includes honoring the fiduciary duties that underlie the doctor's relationship with his or her patients. Delaying medical assistance or neglecting to tend to injuries is not morally acceptable unless a triage emergency requires others to be treated first.

9.4.5 Do Not Disable

Torture is one form of disabling another person. As Gert states, disabilities can be physical, mental, or volitional. Torture generally affects all three, with the infliction of both mental and physical pain. This includes such actions as environmental manipulation, pulling out fingernails, exploitation of fears and phobias, force-feeding, and use of stress positions like shackling.

9.4.6 Do Not Cheat

Cheating is one form of dishonesty often involving methods that give one person an (unfair) advantage over another. Misrepresentation, outright lying, and falsification of documents could be seen as types of cheating. Participating in any of these violate doctors' professional duties and are ethically impermissible.

9.4.7 Do Not Deprive of Freedom

Indefinite detention, long-term imprisonment, solitary confinement, and confinement in small boxes or metal containers are some of the more extreme forms of the deprivation of freedom. All have potentially long-term mental and physical aspects and should warrant the attention of medical personnel. Gert would support doctors speaking out and working together to bring international attention to the harms of these practices. By no means should they stay silent; they have a duty to oppose such deprivation of freedom.

9.4.8 Obey the Law

Torture is explicitly prohibited by international law and any number of professional codes, as will be discussed later in this chapter. By no means should doctors cross that boundary and should be held responsible if they do.

9.4.9 Do Not Deprive of Pleasure

There is no pleasure in being subjected to abusive and degrading treatment, much less torture. Whenever possible, doctors should intervene to stop torture and do all they can to treat injuries and provide comfort for the victims. Even relatively simple pleasures affect the quality of life. For detainees and prisoners such basics as food, shelter, clothing, and sanitary facilities are significant. Detainees have been deprived of solid food from days to months and are often subjected to sleep deprivation.

Many have had to endure forced nudity, prolonged diapering, sexual harassment, and other indignities. Detainee Abu Zubaydah said he was kept seated in a chair and only provided with Ensure liquid and water for 2–3 weeks. Others said they were kept for days to weeks without solid food. Forced hooding has been a common fixture at many detention centers and prisons as well—clearly unpleasant.

9.4.10 Do Your Duty

Doctors have professional duties that guide their behavior and set limits with regard to their relationships with colleagues and patients. Moreover, they should treat their patients with dignity and respect and be attentive to their fiduciary duties to provide adequate health care.

As we can see, Gert's moral duties provide valuable guidance to doctors regarding *their* professional duties and responsibilities. Let's now get an overview of some expectations of major professional organizations regarding health professionals and torture. We will look at the expectations of the following: the World Medical Association and the Declaration of Tokyo, the American Medical Association, the International Council of Nurses, the Istanbul Protocol, and Physicians for Human Rights.

9.5 The World Medical Association

The World Medical Association (WMA) takes an uncompromising stand against torture and roots their position in terms of human rights. Specifically, it puts forward these key points (WMA 2018a, b):

> Torture is one of the most serious violations of a person's fundamental human rights. It destroys dignity, body and mind and has far-reaching effects on family and community.
>
> Freedom from torture is a universal and fundamental human right for all as guaranteed under international law and defined in the UN Convention Against Torture. However, its practice remains alarmingly widespread, particularly in places out of public view.
>
> The WMA unequivocally condemns any involvement of physicians in acts of torture, whether active or passive, as a severe infringement of the International Code of Ethics and human rights law (WMA 2018a, b).

9.5 The World Medical Association

This is in line with the United Nations' General Assembly Principle 2, adopted on 18 December 1982, setting out an expectation for all medical professionals:

> It is a gross contravention of medical ethics … for health personnel, particularly physicians, to engage, actively or passively, in acts which constitute participation in, complicity in, incitement to or attempts to commit torture or other cruel, inhuman or degrading treatment (Human Rights Library 2018).

To summarize, the WMA considers torture and other cruel and degrading acts to be an explicit violation of human rights. It is significant that the debilitating effect on human dignity as well as harm to the victim's family and community was also brought to the fore.

The impact of torture extends far beyond the level of the individual. As a result, any involvement, whether passive or active on the part of health professionals, is to be condemned. It is also important to point out, as does the WMA, how widespread is the practice and how much of it lies outside the public eye.

> The World Medical Association's Declaration of Geneva is a modern restatement of the Hippocratic values. It is a promise by which doctors undertake to make the health of their patients their primary consideration and vow to devote themselves to the service of humanity with conscience and dignity (United Nations sec 60).

This connects to Gert's duty "do not deprive of freedom." Liberty rights should guide us in our ethical decision-making with regard to how detainees and prisoners are treated and the role doctors can play in making and sustaining just policies and practices. We saw this with the opposition of Israeli doctors in 2015 to force-feeding.

Specifically, a new law in Israel requiring physicians to force-feed hunger striking inmates was shunned by medical community as torture, reports *Al-Jazeera*.

> To keep hunger strikers alive and avoid potential public unrest, Israel's parliament, the Knesset, passed the controversial "Law to Prevent Harm Caused by Hunger Strikers" on July 30 [2015], granting authorities the legal right to force-feed prisoners on hunger strike when their lives are in danger.
>
> Members of the IMA [Israeli Medical Association], however, have refused to comply, saying that doing so would be a breach of their Hippocratic Oath. "Forced-feeding is equivalent to torture and every physician has the right to refuse to force-feed a hunger striker against his or her will," the IMA states. (Zahriyeh 2015).

That means both the patient and the doctor have the right to refuse force-feeding. Competent adults have the right to refuse and the right to withdraw from force-feeding. Conscientious medical professionals have the right to refuse initiating or assisting forced feeding.

Furthermore, justice dictates respect for persons and protection from harm—not to mention honoring self-determination. To look at another example, we might note the "Respect for Persons" principle spelled out in the U.S. Navy's guideline regarding human experimentation. It is as follows:

> Respect for Persons: The rights, welfare, interests, privacy, confidentiality, and safety of human subjects shall be held paramount at all times and all research projects shall be conducted in a manner that avoids all unnecessary physical or mental discomfort, and economic, social, or cultural harm (Secretary of the Navy 2006).

This guideline reflects the values that underscore Rawls's attempt to universalize human rights. It is also supported by Gert's duties #1, 2, 3, 5,7, 8, and 10– do not kill, do not deceive, do not cause pain, do not deprive of freedom, obey the law, and do your duty. Each one of these duties support respect for persons and fundamental human rights.

In addition, the Navy's guideline calls for respect for privacy and confidentiality, as well as avoidance of economic, social, or cultural harm. These latter duties are important to point out, given the sorts of problems that have arisen. For example, doctors and psychologists have violated privacy by the use of patient information in developing interrogation methods. Subsequently, phobias such as fear of insects, fear of snarling dogs, or fear of the dark are incorporated in interrogations and other encounters with detainees. In addition, cultural harm has regularly occurred with disrespecting religion and the mistreatment of sacred texts such as the Bible and the Koran.

9.6 The Declaration of Tokyo

The WMA Declaration of Tokyo was adopted in 1975 and most recently revised in 2016. It sets out guidelines for physicians concerning torture and other cruel, inhuman or degrading treatment or punishment of detainees and prisoners. Of utmost significance and influence, the guidelines in abbreviated form are as follows:

WMA Declaration of Tokyo's Key Points

1. The physician shall not countenance, condone or participate in torture or other forms of cruel, inhuman or degrading procedures, whatever the offense and whatever the victim's beliefs or motives.
2. The physician shall not provide any assistance or knowledge to facilitate the practice of torture or other forms of cruel, inhuman or degrading treatment.
3. When providing medical assistance to detainees or prisoners who could later be under interrogation, physicians should ensure confidentiality of all personal medical information.
4. Physicians have the ethical obligation to report abuses, where possible with the subject's consent, but in certain circumstances where the victim cannot express him/herself freely, without explicit consent.
5. The physician shall not use nor allow the use of medical knowledge or skills, or an individual's health information to aid any interrogation, legal or illegal, of those individuals.
6. The physician shall not be present during any procedure during which torture or any other forms of cruel, inhuman or degrading treatment is used or threatened.
7. A physician must have complete clinical independence in deciding upon the care of a person for whom he or she is medically responsible. No motive shall prevail against this higher purpose.

8. Where a prisoner refuses nourishment and is considered by the physician as capable of forming an unimpaired and rational judgment concerning the consequences of refusing nourishment, he or she shall not be fed artificially.
9. The World Medical Association encourages the international community, the National Medical Associations and fellow physicians to support the physician and his or her family in the face of threats or reprisals resulting from a refusal to condone the use of torture or other forms of cruel, inhuman or degrading treatment.
10. The World Medical Association calls on National Medical Associations to encourage physicians to continue their professional development training and education in human rights.

9.7 The American Medical Association

The position of the American Medical Association (AMA) is congruent with that of the WMA. It also sets out a clear statement opposing the participation of doctors in torture, whether actively or passively. Opinion 9.75 of the AMA code asserts the following ethical principles:

1. Physicians must oppose and must not participate in torture for any reason.
2. Physicians must not be present when torture is used or threatened.
3. Physicians may treat prisoners or detainees if doing so is in their best interest, but physicians should not treat individuals to verify their health so that torture can begin or continue.
4. Physicians who treat torture victims should not be persecuted.
5. Physicians should help provide support for victims of torture and, whenever possible, strive to change situations in which torture is practiced or the potential for torture is great.

It is noteworthy that the AMA sets forth a moral duty on the part of the physician to oppose torture and not participate in any way. It is also important to prohibit doctors from even being present as a bystander when such brutality is enacted or threatened. This guideline has the effect of eliminating any question of complicity with what is transpiring. And, we may infer, staying silent is not an option.

Note also that the AMA allows physicians to treat torture victims, but only if it's not in the service of allowing the torture to continue. That frees physicians to put their skills to the service of beneficence and non-maleficence. However, it assumes physicians can make that determination—which may or may not be possible. Here's where the command to "Do no harm" comes in: If it's unclear to the physician how to do good in the context of torture, at least do no harm. If you can use your medical expertise to benefit the torture victim that is good—if not, do no harm. Ideally, it would be advisable to have an additional guideline specifying what path to take if it is unclear if the torture has come to an end or if more lies ahead.

Thirdly, it is commendable that the AMA seeks to protect physicians who care for torture victims so they are not persecuted for lending their help. This latter duty can create challenges, particularly if treating torture victims could be interpreted as complicity. Therefore, it is vital for physicians who do treat torture victims to make their opposition known and take steps to bring the practice to light.

"The physician's most important role is that of healer, and that role is seriously compromised in situations of torture and coercive interrogation," asserted AMA President Robert M. Wah in December 2014, criticizing doctors who assisted the CIA in torturing suspects (Viebeck 2014).

9.8 The International Council of Nurses

The International Council of Nurses (ICN) sets out a similar expectation—and one couched in duty—for the Nursing profession. Specifically:

- ICN strongly affirms that nurses should play no voluntary role in any deliberate infliction of physical or mental suffering.
- The nurse's primary responsibility is to those people who require nursing care.
- Nurses have a duty to provide the highest possible level of care to victims of torture and other forms of cruel, degrading and inhumane treatment, and should speak up against and oppose any deliberate infliction of pain and suffering (1989, rev. 1998).

As far as guidance goes, ICN seeks to provide access to confidential advice and support in caring for prisoners subjected to torture. Also, all levels of Nursing education curricula should recognize human rights issues and violations, such as torture. ICN's position is that:

> Violations of human rights are pervasive and scientific advances have brought about sophisticated forms of torture. Nurses are sometimes called upon to perform physical examinations before prisoners' interrogation and torture, to attend torture sessions in order to provide care, and/or to treat the physical effects of torture (Human Rights Library 1998).

The nurse's professional and ethical duty regarding torture is to the care of victims, but not only that. They also have a duty to let their opposition be known by speaking up and taking action. This latter duty should be explicit, given the prevalence of silence on the part of doctors and nurses against torture.

Commentators observe that doctors and nurses provide credibility for the policies and practices in which they take part. We should hope that this results in their internalizing the two principles of non-maleficence and beneficence and play a leadership role in seeking justice. When others are looking up to them for guidance, they need to make sure that guidance is morally defensible.

9.9 Istanbul Protocol: Obligations to Prevent Torture

The Istanbul Protocol, also known as *The Manual on Effective Investigation and Documentation of Torture and Other Cruel, Inhuman, or Degrading Treatment or Punishment*, was developed in 1999. As noted by Physicians for Human Rights (PHR),

> [It] is the first set of international guidelines for documentation of torture and its consequences. The Istanbul Protocol provides a set of guidelines for the assessment of persons who allege torture and ill treatment, for investigating cases of alleged torture, and for reporting such findings to the judiciary and any other investigative body (2017a, b).

Stated in the Istanbul Protocol are the obligations that States must respect in order to ensure protection against torture. These include: "Taking effective legislative, administrative, judicial or other measures to prevent acts of torture. No exceptions, including war, may be invoked as justification for torture" (United Nations 2004, sec. 10).

The Istanbul Protocol affirms the Hippocratic Oath's emphasis on non-maleficence. This is made explicit by the statement that Western medical values have been dominated by the Hippocratic Oath and similar pledges. In particular, "The Hippocratic Oath represents a solemn promise of solidarity with other doctors and a commitment to benefit and care for patients while avoiding harming them" (United Nations sec. 60). This spotlights three values, those of solidarity, beneficence, and non-maleficence. This links to Beauchamp's and Childress's four basic principles of Western Bioethics; thereby reinforcing their importance.

The manual also contains a promise to maintain confidentiality. This underscores the significance of patient autonomy, as well as the Hippocratic Oath's duty: "Whatever, in connection with my professional practice or not, in connection with it, I see or hear, in the life of men, which ought not to be spoken of abroad, I will not divulge, as reckoning that all such should be kept secret." The Istanbul Protocol asserts:

> These four concepts are reflected in various forms in all modern professional codes of health-care ethics. The World Medical Association's Declaration of Geneva is a modern restatement of the Hippocratic values. It is a promise by which doctors undertake to make the health of their patients their primary consideration and vow to devote themselves to the service of humanity with conscience and dignity. (United Nations sec. 60)

Doctors should keep in mind that there is a strong connection between concepts of human rights and healthcare ethics. The central tenet is "the fundamental duty always to act in the best interests of the patient, regardless of other constraints, pressures or contractual obligations." What follows from this is that "All health professionals … are judged to be guilty of misconduct if they deviate from professional standards without reasonable justification" (United Nations sec. 51).

Under no circumstances is enabling torture a "reasonable justification" for abandoning their professional ethics and, therefore, being guilty of misconduct. Health professionals should never lose sight of this.

9.10 Physicians for Human Rights

Physicians for Human Rights (PHR) has been very active in working for change and directing doctors to use their medical knowledge and skills to bring acts of torture to light. As PHR (2017a, b) notes in its section on anti-global torture,

> Despite the absolute prohibition of torture in international law, it continues to be practiced in more than 100 countries, from totalitarian regimes to democracies. Countries frequently justify the use of torture as a necessary means to extract confessions, identify terrorists, and obtain intelligence critical to preventing future violence. Convictions are difficult to achieve because torturers have become adept at inflicting suffering through methods that leave few physical marks.

Doctors and nurses can play a vital role. In uncovering the systematic use of torture. Physicians for Human Rights points out, "Health professionals can detect signs of physical and mental abuse that are not evident to traditional investigators" (2017a, b). They can then help expose attempts to hide evidence of brutality and, with that, the ethical travesty that was committed (2017a, b).

9.11 Conclusion

We can see from the various moral principles set out by ethicists and international codes of medical ethics that the message is clear. Do not enable, condone, participate in or otherwise support torture or the degrading treatment of others. Standing by the wings while others commit abusive practices is morally repulsive and nothing any conscientious doctor or nurse should have any part in.

In terms of recommendations, strong sanctions and guidelines for reporting abuses need to be in place to underscore the prohibition. Not only do health professionals need to have it clear in their minds that mistreatment in any form is unacceptable, they must also avoid such complicity as falsifying records or other forms of deception.

Medical personnel may be coerced to play a role or provide some legitimacy for torture; for example in a detention facility or out in the field. In such cases, there needs to be channels in place so doctors and other health care givers are not at risk in opposing actions and policies that involve abuse or abdicating their fiduciary duties to their patients. And it needs to be understood that detainees and suspected terrorists are their patients.

Bioethicist Jonathan H. Marks suggests the burden of ethical decision-making not just rest on doctors' shoulders. "Physicians should not be left entirely on their own to chart a course between these competing concerns," he says. "A map should be drawn up with input from military and civilian personnel, physicians, lawyers, ethicists, and laity" (Marks 2005).

One proposal is to make reporting mandatory, as with child abuse. Personnel and management of detention centers would then face penalties for failing to comply (Smith and Freeman 2005, 331). As Smith and Freeman point out, "Those who perpetrate torture rely upon the silence and complicity of their colleagues" (332). In this way the net of culpability is widened to recognize all those who have a role to play, even if only as an enabler. Complicity in any form would then be scrutinized and assessed.

9.12 What, Then, Follows from This?

Perhaps sanctions or expulsion could be considered. Marks (2005) points out examples of professional organizations taking action against doctors enabling torture by setting down sanctions. These include those expelled by the Chilean Medical Society (for overseeing torture under Pinochet), a Brazilian medical council (for falsifying a death certificate of a torture victim), the National Medical Council of Uruguay (for abetting torture under a military junta), to South African sanctions (for failing to report or treat torture of civil rights leader Steve Biko). Such sanctions may be an effective way for professional organizations to bring attention to the issue and discourage further involvement of doctors in enabling torture.

Steven H. Miles notes the increase in holding doctors accountable for torture. He points out:

> Since 1975, the pace of punishing physicians has accelerated. During the 1980s, four countries (Argentina, Chile, South Africa, Uruguay) held physicians accountable for torture. As those four pursued new prosecutions, Brazil and Rwanda joined their ranks during the 1990s. Pakistan and Sri Lanka punished physicians in the first decade of the twentieth century. Since 2010, three more (Guyana, Italy, United Kingdom) have joined them … In the last thirteen years alone, 36 physicians have been punished for crimes against humanity, more than in the years 1975 to 2000 combined (2014).

See also the guidelines of the United Nations Office of the High Commissioner (2018). Physicians for Human Rights' Alan Donaghue calls for psychologists and physicians involved to permanently lose their license. Peter Hall and David N. Tornberg concur that stiff sanctions are in order. They recommend that, "Every one of the 139 countries that ratified the UN Convention against Torture is obligated to either prosecute any suspects entering its territory or extradite them to a country that will" (2005, 1263).

9.13 Working for Change

We need to support organizations like Physicians For Human Rights that focus on ways physicians can work together to oppose torture and know what to do when the evidence suggests that individuals have been victims of torture.

It's one thing to be aware of the ethical codes of one's profession but quite another to hold those codes—honor them— as a moral duty. The work of Physicians for Human Rights, the United Nations, and other professional organizations provide testimony to the importance of working together to effect change. Collaboration matters. It can be a powerful vehicle for bringing problems to light and making an affirmative difference for all concerned.

The knowledge that over 100 countries in the world are guilty of practicing torture indicates that considerable change is in order. The professional perspectives, as we've seen in this chapter, make clear the importance of opposing any form of degrading treatment and abuse of other human beings. They also make it clear that doctors cannot just abstain from participation in torture. They should take an affirmative role in bringing torture to light and exposing those guilty of misconduct.

In light of the principles and the professional duties we have examined in this chapter, we can conclude that there is no excuse for inaction on the part of health professionals.

Works Cited

American Medical Association. Code of Medical Ethics Opinion 9.7.5. https://www.ama-assn.org/delivering-care/torture. Retrieved 10 June 2018.

Beauchamp, Tom L., and James F. Childress. 1979. *Principles of Biomedical Ethics*. Oxford: Oxford University Press.

Beck, Julie. 2014, December 12. 'Do No Harm': When Doctors Torture. *The Atlantic*. https://www.theatlantic.com/health/archive/2014/12/do-no-harm-when-doctors-torture/383677/. Retrieved 10 June 2018.

Beckhusen, Robert. 2012, July 11. U.S. Injected Gitmo Detainees with 'Mind Altering' Drugs. *Wired*. https://www.wired.com/2012/07/gitmo/. Retrieved 10 June 2018.

Gert, Bernard. 2014. In *Global bioethics and human rights: Contemporary perspectives*, ed. Wanda Teays, John-Stewart Gordon, and Alison Dundes Renteln. Lanham MD: Rowman & Littlefield.

Hall, Peter and David N. Tornberg. 2005, October 8. A Stain on Medical Ethics [Letter to the editor]. *Lancet* 366: 1263. http://download.thelancet.com/pdfs/journals/lancet/PIIS0140673605675204.pdf. Retrieved 10 June 2018.

Human Rights Library, University of Minnesota. 1998. Links, ICN Position: International Council of Nurses; Torture, Death Penalty and Participation by Nurses in Executions (1998). http://hrlibrary.umn.edu/instree/executions.html. Retrieved 10 June 2018.

———. 2018. Principles of Medical Ethics Relevant to the Role of Health Personnel. http://hrlibrary.umn.edu/instree/h3pmerhp.htm. Retrieved 10 June 2018.

Istanbul Protocol. Manual on the Effective Investigation and Documentation of Torture and Other Cruel, Inhuman or Degrading Treatment or Punishment. http://www.achpr.org/instruments/istanbul-protocol/. Retrieved 10 June 2018.

Marks, Jonathan H. 2005, July–August. Doctors of Interrogation. *The Hastings Center Report* 35 (4): 17–22. http://www.jstor.org/stable/3528822. Retrieved 10 June 2018.

Miles, Steven H. 2014, January 22. The New Accountability for Doctors Who Torture. *Health and Human Rights Journal*. https://www.hhrjournal.org/2014/01/the-new-accountability-for-doctors-who-utorture/. Retrieved 10 June 2018.

Physicians for Human Rights. 2017a. Documenting Torture Internationally. *Physicians for Human Rights*, PHR.org. http://physiciansforhumanrights.org/issues/torture/international-torture.html. Retrieved 10 June 2018.

———. 2017b. Torture. *Physicians for Human Rights*, PHR.org. http://physiciansforhumanrights.org/issues/torture/. Retrieved 10 June 2018.

Pomerleau, Wayne P. 2018. Western Theories of Justice. *Internet Encyclopedia of Philosophy*. https://www.iep.utm.edu/justwest/. Retrieved 10 June 2018.

Secretary of the Navy. 2006, November 6. SECNAV Instruction 3900.39d, Human Research Protection Program. https://fas.org/irp/doddir/navy/secnavinst/3900_39d.pdf. Retrieved 10 June 2018.

Smith, Henry Forbes, and Mark Freeman. 2005, February 1. The Mandatory Reporting of Torture by Detention Center Officials: An Original Proposal. *Human Rights Quarterly* 27 (1): 327–349. https://muse.jhu.edu/article/178250. Retrieved 10 June 2018.

The Economist. 2004, March 18. Hell-hole or Paradise? https://www.economist.com/united-states/2004/03/18/hell-hole-or-paradise. Retrieved 10 June 2018.

United Nations. 2004. *Istanbul Protocol*. https://www.ohchr.org/Documents/Publications/training8Rev1en.pdf. Retrieved 10 June 2018.

United Nations Office of the High Commissioner. 2018. Principles of Medical Ethics relevant to the Role of Health Personnel, particularly Physicians, in the Protection of Prisoners and Detainees against Torture and Other Cruel, Inhuman or Degrading Treatment or Punishment. https://linkprotect.cudasvc.com/url?a=https%3a%2f%2fwww.ohchr.org%2fen%2fProfessionalinterest%2fPages%2fmedicalethics.aspx&c=E,1,6La-wQaS_BYMtb_15OtNqOwd2i_L4Ba8Tj0yAKeh8_r_beRHUe6XZi0uUWXPJKu61MnRtb2jQaMuX9d5JefG2eiBDK-B06h4jYGIl1B-_lL504w,,&typo=1. Retrieved 10 June 2018.

Viebeck, Elise. 2014, December 12. AMA Rebukes Doctors for Role in CIA 'Torture'. *The Hill*. https://www.google.com/amp/thehill.com/policy/healthcare/227005-ama-rebukes-doctors-for-role-in-cia-torture%3famp. Retrieved 10 June 2018.

World Medical Association. 2018a. Torture Prevention. *World Medical Association*. https://www.wma.net/what-we-do/human-rights/torture-prevention/. Retrieved 10 June 2018.

———. 2018b, April 18. WMA Resolution on the Responsibility of Physicians in the Documentation and Denunciation of Acts of Torture or Cruel or Inhuman or Degrading Treatment. *World Medical Association*. https://www.wma.net/policies-post/wma-resolution-on-the-responsibility-of-physicians-in-the-documentation-and-denunciation-of-acts-of-torture-or-cruel-or-inhuman-or-degrading-treatment/. Retrieved 10 June 2018.

Zahriyeh, Ehab. 2015, August 13. Force- Feeding Palestinian Prisoners Pits Israeli Doctors Against Lawmakers. *Al-Jazeera*. http://america.aljazeera.com/articles/2015/8/13/israeli-doctors-refuse-to-force-feed-palestinian-hunger-strikers.html. Retrieved 10 June 2018.